NATO ASI Series

Advanced Science Institutes Series

A series presenting the results of activities sponsored by the NATO Science Committee, which aims at the dissemination of advanced scientific and technological knowledge, with a view to strengthening links between scientific communities.

The Series is published by an international board of publishers in conjunction with the NATO Scientific Affairs Division

A **Life Sciences** B **Physics**	Plenum Publishing Corporation London and New York
C **Mathematical and Physical Sciences** D **Behavioural and Social Sciences** E **Applied Sciences**	Kluwer Academic Publishers Dordrecht, Boston and London
F **Computer and Systems Sciences** G **Ecological Sciences** H **Cell Biology** I **Global Environmental Change**	Springer-Verlag Berlin Heidelberg New York London Paris Tokyo Hong Kong Barcelona Budapest

PARTNERSHIP SUB-SERIES

1. **Disarmament Technologies**	Kluwer Academic Publishers
2. **Environment**	Springer-Verlag
3. **High Technology**	Kluwer Academic Publishers
4. **Science and Technology Policy**	Kluwer Academic Publishers
5. **Computer Networking**	Kluwer Academic Publishers

The Partnership Sub-Series incorporates activities undertaken in collaboration with NATO's Cooperation Partners, the countries of the CIS and Central and Eastern Europe, in Priority Areas of concern to those countries.

NATO-PCO DATABASE

The electronic index to the NATO ASI Series provides full bibliographical references (with keywords and/or abstracts) to about 50000 contributions from international scientists published in all sections of the NATO ASI Series. Access to the NATO-PCO DATABASE compiled by the NATO Publication Coordination Office is possible in two ways:

- via online FILE 128 (NATO-PCO DATABASE) hosted by ESRIN, Via Galileo Galilei, I-00044 Frascati, Italy.
- via CD-ROM "NATO Science & Technology Disk" with user-friendly retrieval software in English, French and German (© WTV GmbH and DATAWARE Technologies Inc. 1992).

The CD-ROM can be ordered through any member of the Board of Publishers or through NATO-PCO, Overijse, Belgium.

Series I: Global Environmental Change, Vol. 39

Springer
Berlin
Heidelberg
New York
Barcelona
Budapest
Hong Kong
London
Milan
Paris
Santa Clara
Singapore
Tokyo

Microbiology of Atmospheric Trace Gases

Sources, Sinks and Global Change Processes

Edited by

J. Colin Murrell
Department of Biological Sciences
University of Warwick
Coventry CV4 7AL, UK

Donovan P. Kelly
Institute of Education
University of Warwick
Coventry CV4 7AL, UK

Springer
Published in cooperation with NATO Scientific Affairs Division

Proceedings of the NATO Advanced Research Workshop "The Microbiology of Atmospheric Trace Gases: Sources, Sinks and Global Change Processes", held at Il Ciocco, Italy, May 13–18, 1995

Library of Congress Cataloging-in-Publication Data

Microbiology of atmospheric trace gases : sources, sinks and global change processes / edited by J. Colin Murrell, Donovan P. Kelly.
p. cm. -- (NATO ASI series. Series I, Global environmental change ; vol. 39)
Includes bibliographical references and index.

ISBN-13: 978-3-642-64693-5 e-ISBN-13: 978-3-642-61096-7
DOI: 10.1007/978-3-642-61096-7

1. Gases in microorganisms. 2. Biogeochemistry. I. Murrell, J. C. (J. Colin) II. Kelly, Donovan P., 1940- . III. Series.
QR101.M53 1996
576'.15--dc20 95-45667
CIP

ISBN-13: 978-3-642-64693-5 Springer-Verlag Berlin Heidelberg New York

Softcover reprint of the hardcover 1st edition 1996

Typesetting: Camera ready by authors/editors

SPIN: 10498302 31/3137 - 5 4 3 2 1 0

PREFACE

The chapters making up this volume are based on the presentations given by their authors at the NATO Advanced Research Workshop (ARW), also entitled "The Microbiology of Atmospheric Trace Gases: Sources, Sinks and Global Change Processes", held between 13-18 May 1995 at Il Ciocco, Castelvecchio Pascoli, Tuscany, Italy. Four reports of Working Group discussions on aspects of trace gas microbiology and climate change are also included in the volume, prepared by rapporteurs designated at the ARW. All the papers here presented have been subjected to peer review by at least two referees and corrections and amendments made where necessary before their acceptance for publication in this volume.

The ARW was set up to address a wide range of issues relating to atmospheric trace gas microbiology and the organizing group was aware of the burgeoning of studies on gas metabolism and on global effects of atmospheric trace gases over the past two decades. This research effort has led to a number of specialist and generalist meetings including the triennial series of symposia on the metabolism of one-carbon compounds, colloquia concerned with dimethyl sulfide and its precursor, DMSP, through to the Intergovernmental Panels on Climate Change, which have addressed the impact of increasing levels of atmospheric carbon dioxide, methane, nitrous oxide and chlorofluorocarbons on global climate. Over recent years methane and nitrous oxide showed rates of increase in the atmosphere of 40-48 and 3-4.5 Tg/year, respectively. Such increases are microbiologically very significant, as the sources of these gases are almost exclusively from bacterial metabolism, but their increased production frequently results from man's agricultural and land-use practices. Factors affecting sources (and microbiological sinks) for these are thus of global concern for analysis by microbiologists. Other gases of known climatic impact are dimethyl sulfide, and other biogenic sulfur gases such as carbonyl sulfide. Oxidation of these in the atmosphere produces cloud condensation nuclei which can stimulate cloud cover. Natural halocarbons and man-made chlorofluorocarbons have impacts both as stratospheric-ozone-reactive substances and as "greenhouse gases", and the identity and activity of possible microbial sinks for these were of obvious relevance to the ARW. To put into comparative perspective the importance of these trace gases as climatic "drivers", one has only to realize that the "global warming potential" (Radiative Forcing Potential) of some other trace gases is much more potent per molecule than that of carbon dioxide. Taking the molar radiative forcing potential of carbon dioxide as 1 relative unit, the values for some important trace gases are: methane, 21, nitrous oxide, 206, CFC-12 and CFC-113, 15,800 each, HCFC-22, 10,700, and HFC-134a, 9570. All of these compounds are likely to be susceptible to biodegradation, in some cases by bacteria that also oxidize methane, so comparative biochemistry and study of co-metabolism could be of value in such studies. Biodegradation of some of these was considered during the ARW and is reported in this volume.

The development of suitably sensitive analytical instrumentation over the past three decades has enabled precise demonstration of significant concentrations and turnover of a range of atmospheric trace gases, including methane, halomethanes, carbon monoxide, various organosulfur compounds, and the oxides of nitrogen. These are now recognised as major components in the global biogeochemical cycling of carbon, sulfur and nitrogen, with a significant and growing role in the radiative energy balance of the atmosphere (although carbon dioxide and water vapour still currently account for about 25% and 60-70% of the total atmospheric warming globally), and in the dynamics of atmospheric oxidation processes. The seminal role of microorganisms in their biogenesis and ultimate mineralization has only more recently become better documented.

It was a purpose of this ARW to bring together experts in the microbiology and biogeochemistry of these fundamentally important gases, in order to quantify as far as possible the biological driving forces and regulatory processes (sources, sinks, and turnover dynamics) leading to the observed atmospheric composition, and to identify the biological and chemical networks linking the various trace gases in their production and turnover in the terrestrial and marine environments, which are the ultimate sources and sinks for the atmospheric phases of these compounds. This enabled identification of areas in which understanding is inadequate, and where research efforts must be concentrated. A further outcome of the Workshop was a better understanding of the roles of these gases, and their microbiological driving forces, in climate homeostasis and environmental change. The Workshop was structured around (i) a series of theme lectures, in which leaders in this field presented overview and state-of-the-art papers, including the methodology available to investigate molecular ecology and microscale transformations; (ii) oral and poster contributions on current research by participants; and (iii) a number of Working Groups given the brief of arriving at a series of summary conclusions and recommendations concerning the present status of trace gases in biogeochemistry, the microbial processes (and their regulation) that drive these processes, and the needs for future work better to understand the microbiology and the extent to which global environmental change is potentially likely to be forced by microbiological activity. The Workshop brought together specialists who are familiar with the atmospheric processes (Conrad, Oremland, Kelly), the microbiology and ecological dynamics of gas production and consumption in wetlands, coastal and oceanic habitats (Hansen, Iversen, Kiene, King, Lloyd, Nedwell, Panikov, Visscher), and in soil, forest and agricultural systems (Conrad, Iversen, Kelly, Murrell, Steudler). Experimental approaches and analysis of ecological constraints were covered by Kiene, Lidstrom, Murrell and Revsbech. The time was ripe for a synthesis of the kind undertaken, involving participants working on diverse habitats and gases, to integrate understanding (through the Workshop itself and the subsequent publication) of the global processes and the key function of microorganisms in the production and consumption of trace gases, and their role in the catalysis of the complex inter-conversions between them and their precursors. Participants in the ARW are listed at the beginning of the book, and included research students and postdoctoral workers, and thereby extended the scope for trained workers in these fields, as well as providing an expert forum for discussion.

The Editors

CONTENTS

A Global Perspective on Sources and Sinks of Biogenic Trace Gases: an Atmospheric System Driven by Microbiology
D.P. Kelly 1

Physiological Limitations of Methanotrophic Activity *in situ*
G.M. King 17

Methane Production and Oxidation in Soils and Sediments
D.B. Nedwell 33

Methane Oxidation in Coastal Marine Sediments
N. Iversen 51

Microbial Controls of Methane Oxidation in Temperate Forest and Agricultural Soils
P.A. Steudler, R.D. Jones, M.S. Castro, J.M. Melillo and D.L. Lewis 69

Microbial Degradation of Atmospheric Halocarbons
R.S. Oremland 85

Microbial Transformation of Alkyl Esters
D.P. Kelly and A.P. Wood 103

Environmental Molecular Biology Approaches: Promises and Pitfalls
M.E. Lidstrom 121

Molecular Ecology of Methanotrophs
J.C. Murrell, A.J. Holmes, I.R. McDonald and E.M. Kenna 135

Transformation of N_2O and CH_4 in Stratified Microbial Communities Studied by Use of Microsensors
L.R. Damgaard and N.P. Revsbech 153

Metabolism of Nitric Oxide in Soil and Soil Microorganisms and Regulation of Flux into the Atmosphere
R. Conrad 167

Microbiological Controls on Dimethylsulfide Emissions from Wetlands and the Ocean
R.P. Kiene 205

Microbial Turn-over of Volatile Sulfur Compounds
P.T. Visscher 227

Bacterial Metabolism of Methanesulfonic Acid
J.C. Murrell, T. Higgins and D.P. Kelly 243

Working Group Reports:

[1] Contribution of Microbial Processes to Global Budgets
B.H. Svensson 255

[2] Global Environmental Change (GEC)
D.P. Kelly 261

[3] Problems and Developments in Methodology
A. Holmes and D. Lloyd 271

[4] Sulfur, Carbon and Nitrogen Interactions
A.P. Wood 281

Perspectives on the Microbiology of Atmospheric Trace Gases
D.P. Kelly 289

Index 297

Participants at Workshop 301

A Global Perspective on Sources and Sinks of Biogenic Trace Gases: an Atmospheric System Driven by Microbiology

Donovan P Kelly
Institute of Education
University of Warwick
Coventry CV4 7AL
England

Introduction

Atmospheric trace gases containing one or more of carbon, nitrogen and sulfur, in combination with each other or with oxygen, hydrogen or halogens, have long been recognised as key intermediates in the biogeochemical cycling of these elements between the terrestrial and aquatic phases of the biosphere. The pivotal role of microorganisms in the cycling of trace gases was probably first recognized on an international scale by the Symposium held in Göttingen in 1975 (Schlegel et al. 1976). Methane, carbon monoxide, nitrous oxide and dimethyl sulfide occur in the atmosphere only at "trace" levels (ppm or ppb by volume), and until relatively recently it was difficult to measure and monitor their absolute concentrations and dynamics. This served to promote undervaluation of their global importance within the context of biosphere processes. The term "trace" tends to suggest the idea of 'not very much' and hence perhaps "not so important". As Schlegel (1976) put it: "Not long ago these gases were regarded to be curious compounds, far away from the main routes of research and from a quantitative point of view they were considered unimportant. Calculations...culminated in the conclusion that methane is produced in unbelievable amounts on our planet and plays a significant role in the carbon cycle." Subsequently there have been many meetings, many at governmental level, concerned with trace gases, largely because of the growing recognition of the regulatory effect of trace gases, such as methane, the nitrogen oxides, sulfur gases and

NATO ASI Series, Vol. I 39
Microbiology of Atmospheric Trace Gases
Edited by J. Colin Murrell and Donovan P. Kelly

halocarbons, on global climate. The "Greenhouse Effect" has of course been linked particularly to the radiative importance of trace gases other than carbon dioxide and been the subject of dedicated microbiological symposia (Rogers and Whitman 1991).

In the half century since methane was first detected as an atmospheric trace gas (Migeotte 1948), its importance to global climate regulation, and indeed to climate change, through the physicochemical processes which regulate the temperature of the atmosphere has become increasingly apparent. The new techniques for gas analysis which enabled the discovery of methane, carbon disulfide and other trace gases became increasingly sophisticated and sensitive over the decades, allowing the detection of gases present at ever lower concentrations. The study of methane, its very considerable production from biological sources by methanogenic bacteria, and its use as a carbon and energy substrate by methanotrophic bacteria, developed in parallel in the period following the identification of methane as an atmospheric gas. There is now an enormous amount of information about the biological sources and sinks of methane and their regulation, and methane is now known to be the major atmospheric carbon gas after carbon dioxide. Interest in atmospheric trace gases, particularly among climatologists and atmospheric chemists has burgeoned in the last two decades, particularly since the acceptance in general terms of the reality of global warming driven by increasing concentrations of carbon dioxide, methane and nitrous oxide in the atmosphere, largely as direct or indirect results of anthropogenic activity. Nonetheless, the trace gases in general are still the 'orphans' of the biochemical world. While there was enormous development in the understanding of methanogenesis and the biochemistry of methanotrophy during the 1960s and later, understanding of the microbiology and biochemistry of other carbon trace gases did not keep pace with progress of studies on their detection and analysis in the atmosphere, and was until recently restricted to small groups of microbiologists with interests in seemingly esoteric bacteria and their unusual products or substrates. The enormous amount of work done on the anaerobic bacteria reducing either sulfate to hydrogen sulfide or nitrate to nitrogen (and/or nitrous oxide) was for many years largely not correlated with quantitative assessment of the role of those processes in the dynamics and concentrations of those gases on a global atmospheric scale. Only hydrogen sulfide output by sulfate reducing bacteria had been considered on local, global and geological scales in earlier work (Postgate 1979, 1982; Trudinger 1982), and for many years this gas was regarded as the only major input of a sulfur gas into the atmosphere, providing the gaseous link in the sulfur cycle. It is now known that dimethyl sulfide occupies this biogeochemical niche, and is the predominant biogenic sulfur gas, thereby incidentally also adding to the total input of carbon to the atmosphere (see Kiene, Visscher, this volume).

One of the aims of this survey is to promote the status and adjust the perspective of the metabolic trace gases in terms both of quantity of material passing through the 'trace gas' phase and of the diversity of compounds occurring as trace gases, their origins, metabolic and dynamic importance, and their fates in the biosphere.

Carbon dioxide is not here regarded as a 'trace gas' (at around 370 ppm in the atmosphere), making methane (at under 2 ppm) the major biogenic trace gas. Others such as CO, dimethyl sulfide (DMS), nitrous oxide and carbonyl sulfide (COS; the most abundant sulfur trace gas) occur at < 1 ppm steady state levels but all exhibit annual fluxes of one million tonnes (Mt) or more. For methane (in round terms) a steady state concentration of 4000 Mt, and an annual flux of around 630 Mt, represents a major global metabolite, and has both microbial sources and sinks: perhaps 1-10% of all autotrophically fixed carbon dioxide passes through metabolic methane on its way back to carbon dioxide, with at least 1% of all photosynthetically fixed carbon returning to the atmosphere as methane. As with methane and nitrous oxide, the input to the atmosphere of the "lesser" trace gases, DMS, COS, and H_2S (combined output over 100 Mt/year) is the "leakage" from biological cycles turning over vastly greater amounts of these gases in 'semi-closed' oceanic and anaerobic terrestrial habitats than that "leaking" into the atmosphere. In these habitats, 1-10% or more of total biological turnover of carbon and sulfur passes through these compounds.

This overview summarizes the time scales and quantities of gases involved in the turnover of the biogenic carbon, nitrogen and sulfur trace gases (and some also of anthropogenic origin), and to identify the links between the source and sink processes. I have also attempted to identify where future studies need to be focused in order to build a more comprehensive picture of the microbiological dynamics and regulation of gas turnover that might be relevant to modelling global impacts.

Global cycling of carbon

The reservoir sizes of the Earth's carbon and the timescale and size of the fluxes between them, as proposed by Walker (1993) are summarized in Fig. 1. The concerns of the microbiologist are essentially confined to the rapid turnover component of the global system, enclosed within the "box". This model applies of course also to the other biogenic elements found in the trace gases. The importance of microbial processes in forming the geological record and the deduction of the importance of modern processes in many cases relies on deductive analysis of deep sea and geological sediment analysis, including that of sedimentary carbonates and more recent ice cores, both of which preserve records of biological activity involving trace gases in the distant past.

Amounts of trace gases in the atmosphere and their fluxes

Table 1 summarizes the available literature data on the major biogenic gases. Carbon dioxide is included to illustrate the quantitative difference from the true trace gases, and the relative increases of carbon dioxide and methane over the past two centuries (approx. 30% and 40% respectively). While the increases in absolute quantities of these two gases is different by about 75-fold, the increase in the radiative heating effect by the increased methane is about 50% of that caused by the simultaneous increase in the carbon dioxide, emphasizing the importance of understanding the biological and chemical

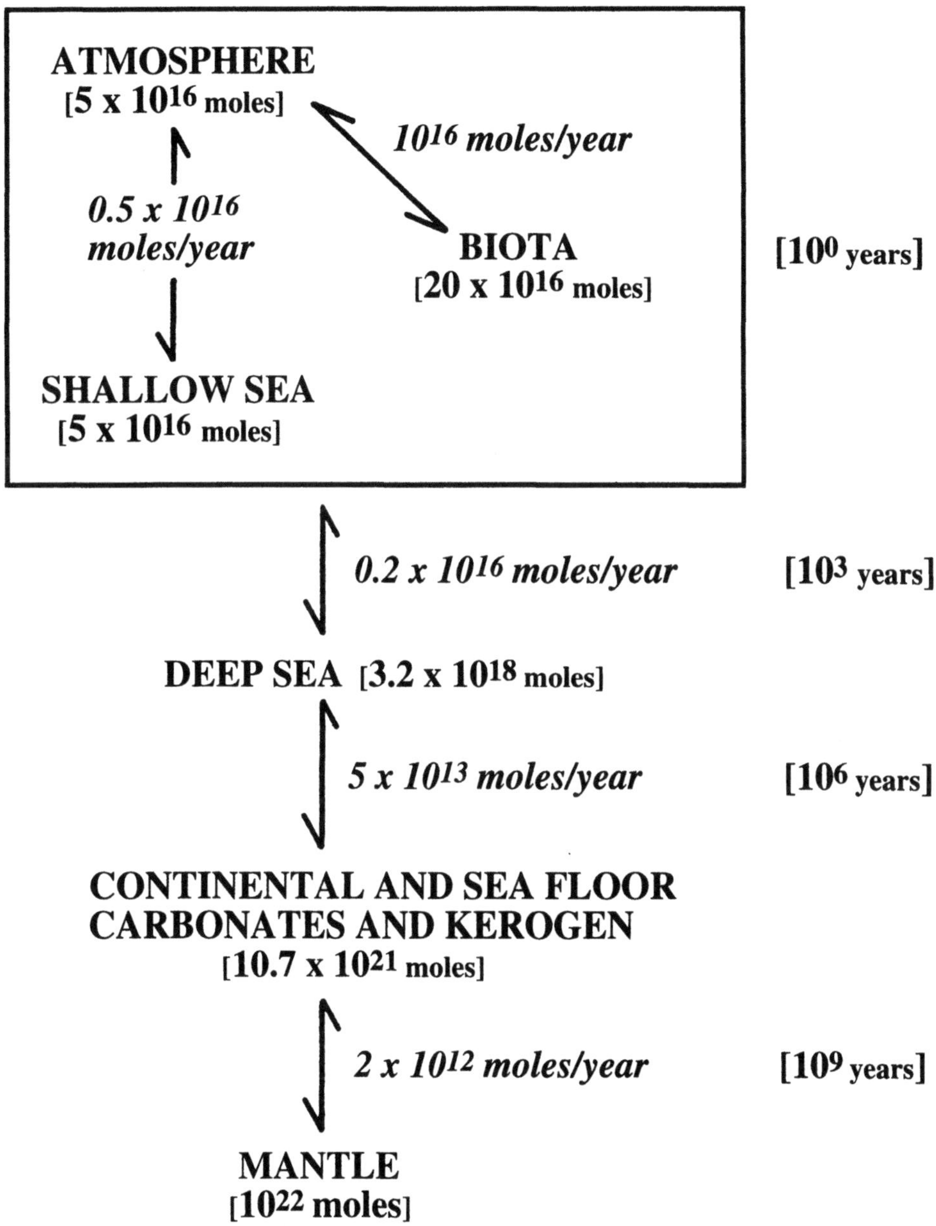

Figure 1. Reservoirs and fluxes between them of carbon on a global and geological time scale (after Walker 1993). Numbers in italics indicate flux rates between reservoirs. Reservoir and flux units: moles of carbon. Numbers at the right of the scheme (powers of 10) indicate overall exchange times between the major reservoirs.

Table 1. A summary compilation of the available literature estimates of the concentrations and fluxes of the principal atmospheric trace gases (see also Kelly and Smith 1990; Kelly et al. 1993, 1994a)

COMPOUND	**PROPORTION**	**ANNUAL PRODUCTION RATE** (10^6 tonnes of compound)	**AMOUNT IN ATMOSPHERE** (10^6 tonnes of compound)	**LIFETIME**
Carbon dioxide	360 ppm [AD1800 = 285]	356,000	2,600,000	10 years
Methane	1.65-1.77 ppm [AD1800 = 0.7]	370-975	4,000-4,600	5-11 years
Carbon monoxide	110 ppb	1,190-1,280 (- 4,400)	550	0.27 year
Hydrogen	500 ppb	36-147	200	2.7 years
Nitrous oxide	310 ppb	9	100	164 years
Dimethyl sulfide	10-100 ppt	40-70	1	1-3 days
Carbonyl sulfide	500 ppt	3	3-4	1 year
Carbon disulfide	1-10 ppt ?	0.57-4.0	?	7-12 days
Methyl chloride	?	3-10	-	-
Ammonia	?	53	-	-
SO_x	?	400	-	-
NO + other NO_x	?	100	-	1 year
Other halocarbons	?	-	-	-
Methanesulfonic acid	?	20-45	-	-
Other CH_3-compounds	?	-	-	-

processes affecting turnover and concentration of atmospheric methane. The relative effects of several trace gases (of both biogenic and anthropogenic origin) on atmospheric radiative forcing are shown in Table 2, and the major sources of some important trace are also outlined (Table 3).

Table 2. Ozone-reactive and greenhouse-effect atmospheric trace gases of anthropogenic origin (McCarty and Reinhard 1991)

COMPOUND	ATMOSPHERIC CONCENTRATION IN 1990 [ppbv]	INCREASE IN RADIATIVE FORCING CAUSED BETWEEN 1980-1990 (%)
Carbon dioxide	353,000	55
Methane	1,720	15
Methyl chloride (0.6 ppbv*)	0.6	0
CCl_4 (0.05)	0.15	-
CH_3CCl_3 (0)	0.17	-
CCl_2FCClF_2 (0)	0.07	7
CCl_3F (CFC-11) (0)	0.26 }	17
CCl_2F_2 (CFC-12) (0)	0.05 }	
Nitrous oxide	310	6

* Numbers in brackets = 1980 concentrations

The detailed processes involving these gases and the nitrogen oxides are the subject of other chapters in this volume and of other specialist reviews (Kelly and Smith 1990; Rogers and Whitman 1991; Kelly et al. 1993; Oremland 1993).

Table 3. A survey of principal sources and sinks for some atmospheric trace gases

COMPOUND	SOURCES	SINKS
METHANE	Sediments, Peat bogs, Ruminant animals Termites Oceans	Soil bacteria Aquatic environments Atmospheric oxidation
DIMETHYL SULFIDE	Oceans: phytoplankton and cyanobacteria Animals, Plants Lichens	Oceans: bacteria Soil bacteria Atmospheric oxidation
CARBONYL SULFIDE	Oceans: chemical CS_2 oxidation Soils (from SCN^-)	Oceans: bacteria? Soil and aquatic bacteria Plants, lichens
CARBON MONOXIDE	Oceans Atmospheric CH_4 oxidation	Soil bacteria
METHYL CHLORIDE	Algae Biomass burning (>50%)	Bacterial oxidation Stratospheric chlorine

Methane in the atmosphere: scale of turnover and dynamics

The amounts of material involved in so-called "trace gas" turnover are on a scale beyond normal experience. Thus, taking the maximum methane output figure from Table 1, the following calculation can be made to put the scale of turnover into a more human dimension: a flux of 975 Mt/year is equivalent to 6.1 x 10^{13} mol/year, equal in volume to 1.365 x 10^{15} liters or 1.365 x 10^{12} cubic meters per year. The "average" domestic household in the UK uses 1500 m^3 of methane annually (as "natural gas"), so if the annual flux into the atmosphere, from all sources (80% biogenic and 20% anthropogenic), could be channelled into that average household, it would supply them for nearly one *billion* years!

Methane undergoes photochemical destruction in the atmosphere by reaction with OH radical, ultimately to produce another trace gas, carbon monoxide. OH and CO, as well as CO + CH_4, are mutual atmospheric sinks, so the sequence of reactions occurring is (in summary):

$$CH_4 + OH \text{ ---> } CH_3 + H_2O \text{ ---[+ OH, } NO_x\text{, light]---> } CO + H_2O$$

$$CO + OH \text{ ---> } CO_2 + H$$

The progressive increase in atmospheric methane in the recent past is still of uncertain cause, and is undoubtedly not explained simply by any one model, as variations in the sources (including diversified global agricultural activity and changes in land use) and sinks (e.g. varying amounts of OH or reactive nitrogen compounds in the troposphere) are not easily modelled into one system. Increase could be due to (i) increases in source emissions to the atmosphere; or (ii) decreased atmospheric destruction; or (iii) a combination of both. Over the period 1962-1987 the increases in atmospheric methane and carbon monoxide were in the ranges 50-65 Mt CH_4 (= 16 ppb or 1.1%) and 4 Mt CO (= 0.8 ppb or 1.1%) per year. Taking combination (iii) as the basis of a time-dependent two dimensional model, Fan et al. (1993) deduced that increase was due to an increase in source emissions of about 10 ppb/year and a decrease in destruction rate of about 6 ppb/year, giving the observed increase of 16 ppb CH_4/year. Their model also indicated that the simultaneous increase in CO (at 0.8 ppb/year) and CH_4 (at 16 ppb/year) was causing a global decrease in atmospheric OH of 0.4% per year. This observation serves to illustrate the interdependence of the atmospheric dynamics of various trace gases, as the OH radical is also a reactant with sulfur gases such as dimethyl sulfide and carbon disulfide in their oxidation to sulfur dioxide, methanesulfonic acid and carbonyl sulfide (see Table 1).

Biological sources and turnover of organosulfur compounds

The principal sources and fluxes of the sulfur gases (listed in Table 1) are shown in Table 4. In addition to these, there are lesser amounts of other biogenic sulfur gases, including methanethiol, dimethyl di- and tri-sulfides, and an array of more complex sulfides and thiophenes produced by plants and animals (Kelly and Smith 1990). Globally, total natural (non-sea-sulfate) sulfur emissions are estimated to be 65 ± 25 Mt Sulfur/year, compared to anthropogenic sulfur emissions of 93 ± 15 Mt Sulfur/year.

Table 4. Natural sources of environmentally significant volatile sulphur compounds (Kelly and Smith 1990; Kelly et al. 1994a; Chin and Davis 1993).

Sources	**Annual output of sulfur gases (Mt Sulfur)**				
	DMS	**CS_2**	**COS**	**H_2S**	**Total**
Oceans	19-51	0.2-0.6	0.3-0.6	<0.5	19.5-52.7
Coastal wetlands	0.6	0.06	0.1	1.0	1.76
Soils and plants	0.2-3.8	0.6-0.8	0.1-0.3	3.2-10	4.1-14.9
Total	19.8-55.4	0.86-1.46	0.5-1.0	4.7-11.5	25.9-69.4

The global inputs of carbonyl sulfide and carbon disulfide from all sources have recently been reassessed (Chin and Davis 1993), and estimated at 1.23 (range 0.83-1.71) Mt COS and 0.57 (0.34-0.82) Mt CS_2 per year, with 30% of atmospheric COS arising from the oxidation of CS_2. The rest of the COS comes from the oceans (28%) and natural terrestrial sources (24%). The major source of CS_2 is believed to be industrial (Kelly and Smith 1990; Chin and Davis 1993), with 58% being anthropogenic. The remainder comes from biological output from aquatic environments, with the greatest output rates from estuaries, less from shelf seas and lowest rates from the open ocean, although the last is the greatest area for emission (Kim and Andreae 1992), and from soils (Kelly and Smith 1990; Kanda et al. 1992; Crozier et al. 1995).

By far the greatest output of organosulfur is in the form of dimethyl sulfide (Tables 1 and 4), and most of this is produced from the breakdown of dimethylsulfoniopropionate (DMSP) produced by marine phytoplankton (Dacey et al. 1994; Matrai and Keller 1994; Wolfe et al. 1994). Seasonal blooms of these can achieve total global areas occupying up to 1.4 million km^2, giving rise to maximum seasonal emission rates to the atmosphere of around 0.7 μmol DMS/m^2/h. This undergoes oxidation in the atmosphere to produce sulfur dioxide and methanesulfonic acid: the latter is a substrate for methylotrophic bacteria after deposition into the oceans or terrestrial environments, completing the organosulfur cycle initiated by dimethyl sulfide synthesis by marine phytoplankton (Kelly et al. 1994b; Prospero et al. 1995; Thompson et al. 1995).

The role of methylated compounds, methylation and demethylation reactions in the production of atmospheric trace gases

Transmethylation to produce volatile compounds and the cleavage of methylated compounds to release gases are both important sources of some trace gases, and it is significant that diverse trace gases are methyl-compounds (DMS, dimethylsulfoxide, methanethiol, methyl chloride, methylated derivatives of mercury, arsenic, selenium and tellurium). Some such reactions (actual and hypothetical) that involve methylation are as follow:

CH_3-DONOR + ACCEPTOR MOLECULE------> CH_3-PRODUCT + OTHER PRODUCT

R-CH_3 + thiol donor --------> methanethiol + [R]
R-CH_3 + chloride -----------> methyl chloride + [R]
R-CH_3 + arsenate -----------> dimethyl arsine + [R]

CH_3-PRECURSOR ---(cleavage)--------> CH_3-VOLATILE

Methane phosphonate + H_2O ----> CH_4 + phosphate
DMSP + H_2O -------------------> DMS + acrylate
Methanesulfonate + H_2O --[?]--> CH_4 + sulfate
Methanesulfonate + H_2O --[?]--> CH_3OH + sulfite

Metabolic and chemical interrelationships among trace gases

Some of these have already been identified, such as the chemical production in the atmosphere of methanesulfonate from dimethyl sulfide, and of carbonyl sulfide from carbon disulfide (and from dimethyl sulfide), as well as the anaerobic interconversions of methylated compounds derived from DMS (Taylor 1993; Visscher, this volume). Methanotrophic or methylotrophic bacteria may remove not only methane (and methanol and methylamines) but some are also able to degrade methanesulfonate, chlorofluorocarbons and alkyl esters (see Oremland, this volume; Kelly and Wood, this volume). The production of nitrous oxide from denitrification can be driven by the oxidation of some methylated compounds, thus linking the carbon and nitrogen trace gases and their precursors. No single gas can thus be studied microbiologically "in isolation" since substrates such as methylated sulfides, methylamines, carbonyl sulfide or chloromethane could all conceivably be metabolized by a single organism, which might also be able to use denitrification as well as aerobic respiration in their dissimilation, thus overlapping the kinetics of turnover of compounds involved in carbon, sulfur, nitrogen and halide cycling.

Final comments

This chapter has provided a scene-setting survey of the trace gases and their place in the global dynamics of the major element cycles. The climatic change effects and other potential impacts on environmental quality of the trace gases, through radiative forcing and ozone-reactivity, are now well-recognized. Microbiological processes are the main sources (and in many cases the ultimate sinks) of many of the key trace gases. Understanding the microbiology and the responses of microbial activity to physical and chemical change in the environment (including interactive responses among microbiologically-driven processes) are key to any modeling of long-term climate change. The overriding importance of carbon dioxide as a regulator of carbon flow in the biosphere (and as a major "greenhouse gas", influencing temperature, another major control on microbial activity) cannot be overemphasised. Recent work shows that rising concentrations of carbon dioxide may greatly enhance CH_4 output from wetlands: raising CO_2 concentrations from a natural 355 ppm to 550 ppm in experimental peat cores more than doubled CH_4 production (Hutchin et al. 1995). This increased bacterial methanogenesis was probably driven by increased input of photosynthate into the habitat by the doubling of the rate of photosynthesis also observed in response to the enhanced CO_2 supply. Such observations make even more crucial a better understanding of the sinks for CO_2 itself, which are still subject to considerable uncertainty, with a "missing sink" for CO_2 as large as 1400 Tg C/year (Houghton and Meira Filho 1995; Schimel 1995).

The subsequent chapters of this book fill in much of the detail, and the unknowns, of the microbiological sources and sinks of all the gases mentioned in this chapter, along with some others. They also serve to provide a basis for assessing the research needs and technique developments required to take the study of the microbiology of atmospheric trace gases to a level at which it can more effectively be applied in the construction of models of global turnover processes. It is clear that major gaps in knowledge include not only poor understanding of the regulatory phenomena determining global production and consumption rates (e.g. of CH_4, DMS and N_2O), but also of the magnitude of the fluxes of even these major trace gases, let alone of those with lower flux rates (CS_2, COS, halocarbons, methylamines and so on). It is obviously necessary to establish all the minor as well as major microbiological contributors to fluxes (e.g. methane consumption by ammonia-oxidizers, sinks for methanesulfonate other than aerobic methylotrophs, nitrifiers as sources of atmospheric nitrogen oxides), since uncertainty about sources and sinks may invalidate a calculated global budget, and will certainly put in doubt any assumptions about environmental factors likely to affect particular fluxes in the future. Uncertainties in the magnitude of the budget for one gas could have drastic effects on assumptions about the fluxes of others: for example uncertainty about methane affects assumptions about dimethyl sulfide, as both are reactants with OH radicals, as is CO, which has natural sources other than OH-dependent oxidation of CH_4. A major residual uncertainty in the methane cycle is the identity of anaerobic methane-oxidizing bacteria or consortia, and their importance as regulators of the amount of methane leaving anaerobic source environments. A very important challenge for microbiology is the continued search for physiological diversity and to expand the portfolio of organisms known to effect reactions that produce or consume the carbon, nitrogen, sulfur and halogenated trace gases.

References

Chin M, Davis DD (1993) Global sources and sinks of OCS and CS_2 and their distributions. Global Biogeochem Cycles 7:321-337

Crozier CR, Devai I, DeLaune RD (1995) Methane and reduced sulfur gas production by fresh and dried wetland soils. Soil Sci Soc Amer J 59:277-284

Dacey JWH, King GM, Lobel PS (1994) Herbivory by reef fishes and the production of dimethylsulfide and acrylic acid. Mar Ecol Prog Ser 112:67-74

Fan SM, Rodriguez JM, Ko MKW, Sze ND (1993) A time-dependent two-dimensional-model study of the trend in atmospheric methane. In Oremland RS (ed) (1993) Biogeochemistry of global change. Radiatively active trace gases, pp 98-112. Chapman and Hall New York London

Houghton JT, Meira Filho LG (1995) The IPCC report on radiative forcing of climate change. Cambridge University Press Cambridge

Hutchin PR, Press MC, Lee JA, Ashenden TW (1995) Elevated concentrations of CO_2 may double methane emissions from mires. Global Change Biol 1:125-128
Kanda K, Tsuruta H, Minami K (1992) Emission of dimethyl sulfide, carbonyl sulfide, and carbon disulfide from paddy fields. Soil Sci Plant Nutr 4:709-716
Kelly DP, Smith NA (1990) Organic sulfur compounds in the environment: biogeochemistry, microbiology, and ecological aspects. Adv Microb Ecol 11:345-385
Kelly DP, Malin G, Wood AP (1993) Microbial transformations and biogeochemical cycling of one-carbon substrates containing sulphur, nitrogen or halogens. In Murrell JC, Kelly DP (eds) Microbial growth on C_1 compounds, pp 47-63. Intercept Andover
Kelly DP, Wood AP, Jordan SL, Padden AN, Gorlenko VM, Dubinina GA (1994a) Biological production and consumption of gaseous organic sulphur compounds. Biochem Soc Trans 22:1011-1015
Kelly DP, Baker SC, Trickett J, Davey M, Murrell JC (1994b) Methanesulphonate utilization by a novel methylotrophic bacterium involves an unusual oxygenase. Microbiology (UK) 140:1419-1426
Kim KH, Andreae MO (1992) Carbon disulfide in the estuarine, coastal, and oceanic environments. Mar Chem 40:179-197
Matrai PA, Keller MD (1994) Total organic sulfur and dimethylsulfoniopropionate in marine phytoplankton: intracellular variations. Mar Biol 119:61-68
McCarty PL, Reinhard M (1991) Biological and chemical transformations of halogenated aliphatic compounds in aquatic and terrestrial environments. In Oremland RS (ed) (1993) Biogeochemistry of global change. Radiatively active trace gases, pp 839-852. Chapman and Hall New York London
Migeotte MV (1948) Spectroscopic evidence of methane in the earth's atmosphere. Phys Rev 73:519-520
Oremland RS (ed) (1993) Biogeochemistry of global change. Radiatively active trace gases. Chapman and Hall New York London
Postgate JR (1979) The sulphate reducing bacteria. Cambridge University Press
Postgate JR (1982) Economic importance of sulphur bacteria. In Postgate JR, Kelly DP (eds) Sulphur bacteria, pp 153-170. The Royal Society, Cambridge University Press
Prospero JM, Savoie DL, Arimoto R, Olafsson H, Hjartarson H (1995) Sources of aerosol nitrate and non-sea-salt sulfate in the Iceland region. Sci Total Environ 160:181-191
Rogers JE, Whitman WB (1991) Microbial production and consumption of greenhouse gases: methane, nitrogen oxides, and halomethanes. American Society for Microbiology Washington DC
Schimel DS (1995) Terrestrial ecosystems and the carbon cycle. Global Change Biol 1:77-91
Schlegel HG (1976) Introductory remarks. In Schlegel HG, Gottschalk G, Pfennig N (eds) Symposium on microbial production and utilization of gases pp 11-12.Goltze Göttingen
Schlegel HG,Gottschalk G, Pfennig N (eds) (1976) Symposium on microbial production and utilization of gases. Goltze Göttingen
Taylor BF (1993) Bacterial transformations of organic sulfur

compounds in marine environments. In Oremland RS (ed) (1993) Biogeochemistry of global change. Radiatively active trace gases, pp 745-781. Chapman and Hall New York London

Thompson AS, Owens NJP, Murrell JC (1995) Isolation and characterization of methanesulfonic acid-degrading bacteria from the marine environment. Appl Environ Microbiol 61:2388-2393

Trudinger PA (1982) Geological significance of sulphur oxido-reduction by bacteria. In Postgate JR, Kelly DP (eds) Sulphur bacteria, pp 133-151. The Royal Society, Cambridge University Press

Walker JCG (1993) Biogeochemical cycles of carbon on a hierarchy of time scales. In Oremland RS (ed) (1993) Biogeochemistry of global change. Radiatively active trace gases, pp 3-28. Chapman and Hall New York London

Wolfe GV, Sherr EB, Sherr BF (1994) Release and consumption of DMSP from *Emiliana huxleyi* during grazing by *Oxyrrhis marina*. Mar Biol Prog Ser 111:111-119

Physiological Limitations of Methanotrophic Activity *in situ*

Gary M King
Darling Marine Center
University of MaineWalpole
ME 04573
USA

Methanotrophic bacteria play an important role in the atmospheric methane budget that at present includes a biospheric flux of about 400 Tg yr-1 (e.g. Cicerone and Oremland, 1988). The role of methanotrophs is primarily a function of two processes: 1) consumption of atmospheric methane by soils, accounting for about 40 Tg yr-1 (see King, 1992; Reeburgh et al., 1993), and 2) oxidation of methane that diffuses to oxic surfaces in wetlands, especially the sediment-water (or air) interface and the rhizosphere of rooted aquatic vegetation (e.g. King, 1990a; King et al., 1990b; Chanton and Dacey, 1991; Schütz et al., 1991; King, 1994). The magnitude of methane consumption by the latter process is uncertain. However, it may be approximated by making several assumptions: 1) 90% of the total wetland emission term results from methane transport through plants (Chanton and Dacey, 1991 and references therein); 2) 25% of the methane that diffuses into plants and 80% of the methane diffusing to the sediment-water (or air) interface is oxidized (King, 1990b; Epp and Chanton, 1993; King, in prep.). Based on these assumptions and a combined emission term for natural wetlands and rice paddies of 200 Tg yr-1 (Aselmann and Crutzen, 1989), methane oxidation amounts to 80 Tg yr-1 at the sediment-water (or air) interface and 60 Tg yr-1 in the rhizosphere, for a combined total of 140 Tg yr-1 (this is about 41% of the wetland methane budget [production = oxidation + emission]). Total wetlands net primary production (NPP), including that for rice paddies, provides a constraint on the extent of methane oxidation. Using values of 50 and 90% for oxidation, the emission term above, and a 1:1 stoichiometry for CH_4:CO_2 production during the fermentative

NATO ASI Series, Vol. I 39
Microbiology of Atmospheric Trace Gases
Edited by J. Colin Murrell and Donovan P. Kelly

degradation of organic matter, one can calculate that total carbon flow through methanogenesis is respectively about 15 and 75% of the estimated 4000 Tg carbon in NPP (Aselmann and Crutzen, 1989). These estimates of relative carbon flow provide reasonable bounds for the relative extent of methane oxidation.

Though the preceding calculations are based on speculative assumptions, it is interesting to note that atmospheric methane consumption by soils and the two wetland oxidation terms are similar in magnitude and substantially larger than the approximately 4 Tg yr-1 by which the atmospheric methane burden is increasing (Blake and Rowland, 1988; Steele et al. 1992). Thus, small but sustained changes in methanotrophic activity can contribute significantly to changes in atmospheric concentrations.

What, then, regulates methanotrophic activity? Clearly, regulation is a function of a number of factors operating at organismal, ecosystem and biospheric levels. For example, at regional to biospheric scales, patterns of surface water distribution are key determinants of wetland area, the position of the water table in wetlands, and the extent of gas spaces in soils, all of which affect both methane production and methane oxidation (e.g. Aselmann and Crutzen, 1989; Moore and Knowles, 1989; Adamsen and King, 1993; Roulet et al., 1993). The hydrologic cycle, and hence regional patterns of water distribution, are in turn very sensitive to the Earth's radiation balance. Historical shifts in the concentration of atmospheric methane from about 0.3 to 0.7 ppm have been associated with climate changes that have expanded or contracted the areal extent of both northern and tropical wetlands (Chappellaz et al., 1990; Crowley, 1991); these changes have been attributed to global-scale temperature-driven changes in surface hydrology (Chappellaz, 1990; Street-Perrott, 1992). At an ecosystem scale, hydrological phenomena are obviously important controls of methanotrophic activity. However, additional factors are also important. The distribution and turnover of organic nitrogen, especially as these parameters affect ammonium, likely contribute to the distribution of methanotrophic bacteria (Steudler et al., 1989; Mosier et al., 1990; King and Schnell, 1994; Schnell and

King, 1994). Competition among methanotrophic bacteria and between methanotrophs (e.g., Graham et al., 1993) and other microbiota is undoubtedly another key control. The latter likely involves molecular oxygen, a potential limiting factor for both rhizospheric methane oxidation and oxidation at the sediment-water interface (King, 1990b; King, 1994; King, in prep.). Symbiotic interactions between methanotrophs and higher organisms, both plant and animal, as well as other indirect interactions, may partially determine methanotrophic activity in a variety of systems (e.g. Childress et al. 1986; Fischer et al., 1987; Cavanaugh et al. 1987; King, 1994).

The focus of this manuscript is an additional set of controls, specifically those based on microbial physiology. Physiological controls are no more or less important than others. However, the physiology of methanotrophs sets the operational limits for activity in a phase space defined by parameters such as methane and oxygen concentrations, pH, temperature, water potential, and salinity (see reviews by Anthony, 1982; Bédard and Knowles, 1989; King, 1993). Further, the controls acting at ecosystem and biospheric levels are at least partially manifest at a physiological level. Thus, I summarize here aspects of methanotrophic physiology that are potentially important determinants of activity in soils and aquatic systems, with an emphasis on the responses of methanotrophs to typical methane concentrations in situ, and to the absence of oxygen.

Are methane-oxidizing bacteria oligotrophs?

Methanotrophs are typically cultured with gas phase methane concentrations between 10-50% (vol/vol), resulting in initial dissolved methane concentrations between about 0.14-0.7 mM. However, methane concentrations in situ are generally low nano- to micromolar. For instance, dissolved methane in the photic zone of the ocean is < 10 nM (e.g., Griffiths et al., 1982; Ward and Kilpatrick, 1990), while values in oxic soils are < 2.5 nM (see for example profiles in Adamsen and King, 1992). These very low concentrations immediately raise questions about the status of methanotrophs vis-a-vis substrate concentrations: are methanotrophs copiotrophs or oligotrophs specifically adapted to regimes of limited substrate availability?

Poindexter (1981) has defined oligotrophic bacteria as those that can be isolated on media containing 0.08-1.25 mmol C 1-1. This range obviously includes concentrations suitable for the culture of all known methanotrophs. Poindexter (1981) has also suggested that environments (and the organisms in them) can be characterized on the basis of a carbon flux term. In aquatic systems, values < about 10 µmol 1-1 d-1 were considered oligotrophic. Methane oxidation in wetlands, lake sediments and rice paddies probably exceeds this "critical" value, often by at least 10- to 100-fold (e.g., Frenzel et al., 1990; King, 1990b; Conrad and Rothfuß, 1991; Sinke et al., 1992; Bosse et al., 1993), though exceptions undoubtedly exist for less productive systems (e.g., Remsen et al., 1989). In contrast, methane oxidation in the water column of many lakes and marine systems can be characterized as oligotrophic (e.g., Rudd and Hamilton, 1978; Harrits and Hanson, 1980; Griffiths et al., 1982; Ward and Kilpatrick, 1990). Likewise, typical rates of atmospheric methane oxidation in soils meet an oligotrophic criterion, though in general these rates appear higher than activity in the water column (e.g., Born et al., 1990; Mosier et al., 1991; Whalen and Reeburgh, 1990; Striegl et al., 1992; Adamsen and King, 1993; Koschorreck and Conrad, 1993; Tate and Striegl, 1993). Thus, the characteristics of a number of systems might select for methanotrophic oligotrophs in addition to the oligotrophic heterotrophs that have been reported (see Kuznestov, 1979; Poindexter, 1981).

A number of adaptations promote the ability of oligotrophic bacteria to successfully utilize substrates present at low concentrations or fluxes (Poindexter, 1981). These include some features that are atypical of the methanotrophs presently in culture. For example, methanotrophs are not known for their high surface to volume ratios, high affinity uptake systems, distinct morphological transitions (e.g., bacillus to coccus), or the utilization of diverse organic substrates (the broad specificity of methane monooxygenases nothwithstanding [Burrows et al., 1984]).

However, it is not at all evident that the methanotrophs in culture are representative of the methanotrophs that are active in situ In fact, typical enrichment and isolation protocols could select against oligotrophic methanotrophs, if, like other oligotrophs, they are unable to utilize growth-supporting substrates at high concentrations (Kuznetsov et al., 1979). Instead, conventional culture methods might select for relatively inactive populations that have a more copiotrophic character. The use of low methane concentrations and long-term enrichments might prove especially fruitful for isolating novel but more representative marine and soil methanotrophs. For example, though a number of marine isolates have been described (e.g., Sieburth et al., 1987; Lidstrom, 1988; Lees et al., 1991), few appear well adapted to the marine environment, particularly with respect to salinity optima, but perhaps also with respect to the utilization of low methane concentrations. Likewise, several lines of evidence (see below) indicate that the methanotrophs isolated to date from soils may not be responsible for atmospheric methane consumption by soils.

Metabolic responses to low methane concentrations

Whether oligotrophic or not, some methanotrophs display a "starvation" response to methane concentrations that typically occur in situ. For cultures of *M. trichosporium* OB3b, a starvation response is apparently invoked at methane concentrations of about 100 p.p.m. or less (Bowman and Sayler, 1994; Roslev and King, 1993; 1994). While both *M. trichosporium* OB3b and *Methylobacter albus* BG8 grown at high methane concentrations can oxidize low concentrations temporarily, neither can grow on atmospheric methane, and both substantially lose the capacity to oxidize low concentrations within 48 hr (Schnell and King, in press) with an accompanying decrease in viability (Roslev and King, 1994).

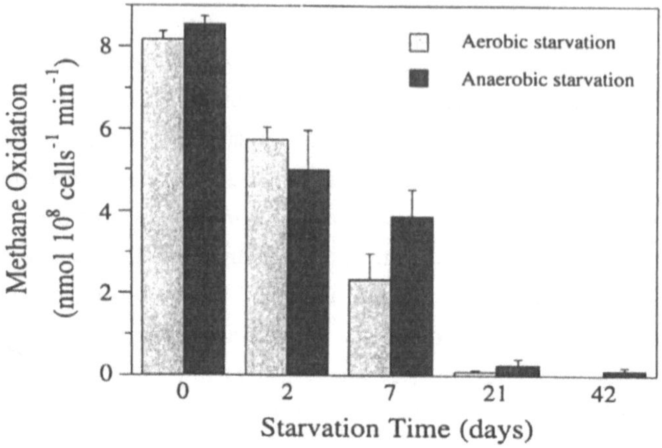

Figure 1. Post-starvation rates of methane oxidation by *M. trichosporium* OB3b starved under aerobic and anaerobic conditions (from Roslev and King, 1994).

The starvation response of *Methylosinus trichosporium* OB3b is manifest as changes in cell number, methane uptake per cell (or per mg protein), naphthalene oxidation, and cell protein, nucleic acid and lipid contents (Bowman and Sayler, 1994; Roslev and King, 1993; 1994). The changes in both protein levels and methane uptake for starved *M. trichosporium* OB3b may be explained by catabolic degradation of methane monooxygenase. The lack of protein loss and higher methane oxidation potential for cells starved under anoxic conditions provides further support for this suggestion (Roslev and King, 1994).

The significance of a starvation response in methanotrophs is not entirely clear, and the relative capacity of methanotrophs versus other heterotrophs to survive under starvation conditions is uncertain. It is clear however that many methanotrophs possess some form of resting stage, either cysts or spores, and that a number of taxa accumulate storage compounds that are often associated with starvation metabolism (Whittenbury et al., 1970a, b; Dawes, 1985; Williams, 1985). These characteristics may reflect the fact that soils, the oceans, and the water column of most lakes are characterized by starvation methane levels. Based on the responses of *M.*

trichosporium OB3b and *M. albus* BG8, only a very small fraction of the Earth's surface, for example, oxic wetland sediments and the rhizosphere of wetland plants, appears conducive for active growth by some of the known

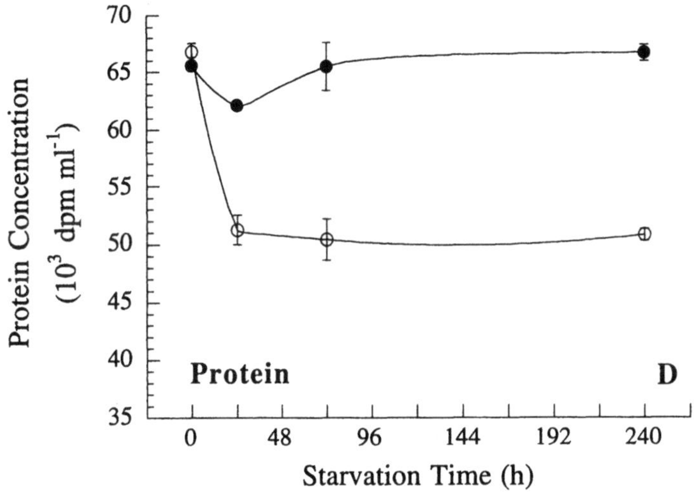

Figure 2. Changes in the radioactivity recovered in a protein fraction during aerobic (O) and anaerobic (●) starvation of *M. trichosporium* OB3b grown with $^{14}CH_4$. From Roslev and King (1995).

methanotrophs. Isolates such as *M. trichosporium* OB3b and *M. albus* BG8 must persist in non-wetland habitats primarily as dormant or resting stages, with active methanotrophy due to other taxa that have lower thresholds for their starvation responses.

Results from soil analyses illustrate contrasting concentration thresholds for starvation. Relatively constant uptake rates can be maintained for long periods (>100 d) when soils are continually exposed to atmospheric methane (Schnell and King, in press). However, soils that are maintained with methane-free air (< 0.03 p.p.m.) rapidly lose their capacity for methane uptake; this capacity is not restored even after prolonged incubation (16 d) with 1.7 p.p.m. methane.

Methanotrophic activity in soils enriched *in vitro* by incubation with 1000 p.p.m. is also lost rapidly after incubation with atmospheric methane. Interestingly, the extent of dimunition with time appears similar for the uptake of atmospheric methane, the activity enriched by incubation with 1000 p.p.m. methane, and the activity of pure cultures of *M. trichosporium* OB3b (Roslev and King, 1994; Schnell and

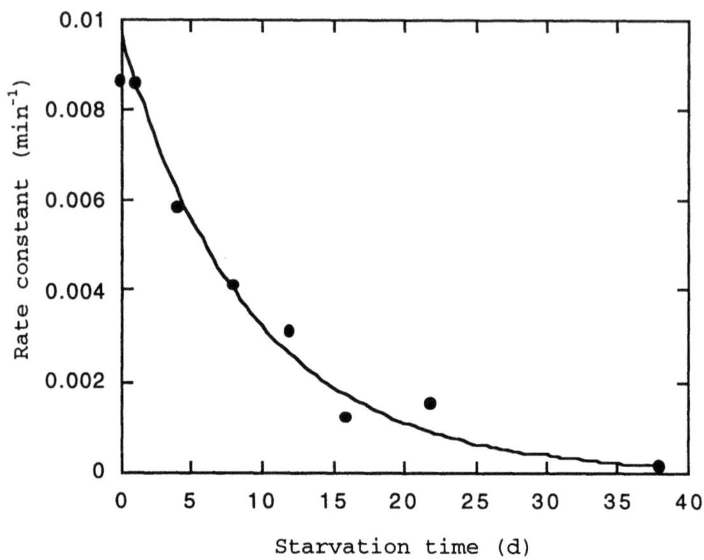

Figure 3. Post-starvation atmospheric methane uptake rates for soils incubated *in vitro* for varied periods with < 0.03 p.p.m. methane in air. From Schnell and King (in press).

King, in press). This suggests that there are physiological similarities between methanotrophs with both high and low methane concentration thresholds for starvation. In addition, the loss of enriched activity upon incubation with atmospheric methane suggests that the populations responding to high methane concentrations differ from those that consume atmospheric levels; this is consistent with observations by Bender and Conrad (1992) of two kinetically distinct groups of soil methanotrophs. The differences in threshold for starvation response may reflect differences in both methane uptake kinetics and maintenance energy requirements. Methanotrophs active in oligotrophic systems may be physiologically similar to other methanotrophs, yet have lower maintenance needs, higher affinities and lower maximal uptake velocities.

The induction of a starvation response in soils incubated for a relatively brief period with methane-free air also indicates that endogenous multi-carbon compounds and non-methane C1 substrates are unimportant in atmospheric methane consumption. Further, the observed starvation response eliminates ammonia-oxidizing bacteria as the primary consumers of atmospheric methane. The methanotrophs responsible for this process appear primarily, if not exclusively, dependent on methane as a source of carbon and energy. This is, of course, consistent with the behavior of methanotrophs in general, and is concordant with the lack of response of atmospheric methane uptake to additions of various potential organic substrates (Schnell and King, in press).

The low concentrations that result in methane starvation in soils are consistent with methane concentrations observed in situ, and can be understood as a survival response in sub-surface horizons where methane drops well below atmospheric levels. The ecological relevance of a relatively high concentration trigger for methane starvation (as observed in *M. trichosporium* OB3b) is less easily understood. A high threshold may simply be characteristic of those methanotrophs that preferentially exploit relatively high methane concentrations, such as those at the sediment-water (air) interface of lakes and wetlands. A comparative analysis of the concentration trigger for starvation responses of a variety of aquatic and soil samples should prove revealing in this context.

Metabolic responses of methanotrophs to anoxia

Although methanotrophic starvation physiology can be logically considered in the context of oxic conditions, it is evident that exposure to anoxic conditions in situ necessitates a metabolic response not typically associated with methanotrophs. The relevance of at least a maintenance-level anaerobic metabolism can be understood by considering the characteristics of those environments where high methane concentrations and molecular oxygen co-occur. At the sediment-water (or air) interface in wetlands, methane and oxygen concentrations are inversely related, with oxygen often limited to mm vertical scales (e.g., King et al., 1990). In these systems, oxygen distributions can fluctuate significantly along a depth gradient as a result of variations in benthic photosynthesis, in some instances causing a given horizon to shift between oxic and anoxic conditions (e.g., King, 1990a; King et al., 1990). In addition, sediment deposition can result in the burial of methanotrophs to depths below which oxygen does not penetrate. Methanotrophs associated with roots and rhizomes are also likely subjected to shifts in the availability of molecular oxygen, especially as a consequence of plant scenescence. The ability of methanotrophs to survive anoxia is amply illustrated by depth profiles that show substantial and immediate methane oxidation by anoxic peats or sediments incubated with oxygen (e.g., King, 1990b; King et al., 1990; Roslev and King, in prep.; see also King, 1992). Even material that has likely been devoid of oxygen for months-years responds rapidly to oxygen re-introduction, too rapid for the germination of cysts or spores to provide an explanation for the observed methanotrophic activity.

Short- and long-term survival of methanotrophs under anoxic conditions both require some form of basal anaerobic metabolism that has been documented only recently (Roslev and King, 1993; 1994; 1995). Results from studies of ^{14}C[UL]-glucose uptake and metabolism, and the distribution of radioactivity in cells incubated with $^{14}CH_4$, indicate that *M. trichosporium* OB3b and an unidentified Group 1 peat isolate are capable of a slow but significant fermentation (Roslev and King, 1995), that in the case of glucose appears to produce acetate by an as yet unknown mechanism. Though it seems highly unlikely that a methanotrophic fermentation could significantly affect total anaerobic metabolism, it is interesting to note that the methanotrophs present in anoxic sediments might contribute to the production of methane that is consumed by other methanotrophs in overlying oxic sediments.

It is also important to note that anaerobic starvation appears less stressful metabolically than aerobic starvation. Methanotrophs show a lower survival rate, lower post-starvation methane oxidation rates and greater losses of protein and lipids under aerobic than anaerobic starvation conditions (Roslev and King, 1994; 1995). This may be due in part to damage by oxygen radicals or peroxides, and the depletion of NADH + H+ due to the oxidase reaction of methane monooxygenase in the absence of methane (i.e., oxygen is reduced to water; Dalton, [1992]). In this context, it appears that some methanotrophs are better adapted for survival under anoxic conditions than under aerobic conditions with low methane concentrations.

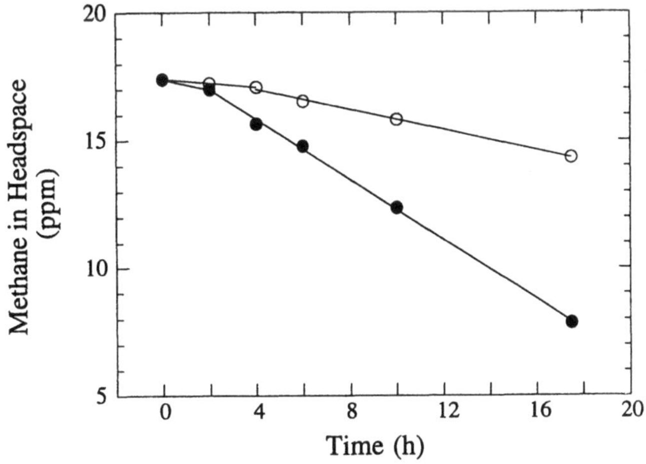

Figure 4. Time course of methane uptake by cultures of *M. trichosporium* OB3b after addition of methane and air; cultures were incubated under oxic (o) or anoxic starvation (•) conditions for 7 d. From Roslev and King (1994).

Summary

The physiological attributes of methane-oxidizing bacteria play a major role in determining the importance of methanotrophy in the global methane budget. The response of methanotrophs to the prevailing concentrations in soils and aquatic systems is of particular interest. Analyses of several well-characterized pure cultures indicate that the nano- to low micromolar methane concentrations that often occur in situ represent starvation levels. Consequently, many of the known methanotrophs may be active in only a few very restricted habitats, while other as yet unisolated methanotrophs dominate soil and water column methane oxidation. Isolation of these organisms might be facilitated by adopting some of the techniques used for enriching oligotrophic bacteria. The ability of methanotrophs to survive both short- and long-term anoxia also deserves additional attention. This capability, along with the capacity of soil methanotrophs to sustain activity with low fluxes of methane, suggests that at least some methanotrophs have developed very low maintenance energy requirements.

Acknowledgements

I thank Drs. S. Schnell and P. Roslev for many helpful comments and insights. This work was supported in part by the National Science Foundation (DEB-9107315), NASA (NAGW-3746) and the U.S. Department of Agriculture (94-37107-0488). Contribution 284 from the Darling Marine Center.

References

Adamsen APS, King GM (1993) Methane consumption in temperate and sub-arctic forest soils: rates, vertical zonation and response to water and nitrogen. Appl Environ Microbiol 9:485-490

Anthony C (1982) The biochemistry of methylotrophs. Academic Press, New York 431 pp

Aselmann I, Crutzen PJ (1989) Global distribution of natural freshwater wetlands and rice paddies, their net primary productivity, seasonality and possible methane missions. J Atmos Chem 8:307-358

Bédard C, Knowles R (1989) Physiology, biochemistry, and specific inhibitors of CH4, NH4+, and CO oxidation by methanotrophs and nitrifiers. Microbiol Rev 53:68-84.

Bender M, Conrad R (1992) Kinetics of CH4 oxidation in oxic soils exposed to ambient air or high CH4 mixing ratios. FEMS Microbiol Ecol 101:261-270

Blake DR, Rowland FS (1988) Continuing worldwide increase in tropospheric methane, 1978-1987. Science 239:1129-1131

Born M, Dörr H, Ingeborg L (1990) Methane consumption in aerated soils of the temperate zone. Tellus 42(B):2-8

Bosse U, Frenzel P, Conrad R (1993) Inhibition of methane oxidation by ammonium in the surface layer of a littoral sediment. FEMS Microbiol Ecol 13:123-134

Bowman JP, Sayler SG (1994) Optimization and maintenance of soluble methane monooxygenase activity in *Methylosinus trichosporium* OB3b. Biodegradation 5:1-10

Burrows KJ, Cornish A, Scott D, Higgins IG (1984). Substrate specificities of the soluble and particulate methane monooxygenases of *Methylosinus trichosporium* OB3b. J Gen Microbiol 130:3327-3333

Cavanaugh CM, Levering PR, Maki JS, Mitchell R, Lidstrom M (1987). Symbiosis of methylotrophic bacteria and deep-sea mussels. Nature (London) 325:346-348

Chanton JP, Dacey JWH (1991). Effects of vegetation on methane flux, reserviors, and carbon isotopic composition. pp. 65-92. In, Sharkey TD, Holland EA, Mooney HA (eds.), Trace gas emissions by plants. Academic Press, Inc., New York

Chappellaz J, Barnola JM, Raynaud D, Korotkevich YS, Lorius C (1990). Ice-core record of atmospheric methane over the past 160,000 years. Nature (Lond.) 345:127-131

Childress JJ, Fisher CR, Brooks JM, Kennicutt MC, Bidigare R, Anderson A. 1986. A methanotrophic marine molluscan (*Bivalvia, Mytilidae*) symbiosis: mussels fueled by gas. Science 233:1306-1308

Cicerone RJ, Oremland RS (1988) Biogeochemical aspects of atmospheric methane. Glob Biogeochem Cyc 2:299-327

Conrad R, Rothfuß, F (1991) Methane oxidation in the soil surface layer of a flooded rice field and the effect of ammonium. Biol Fertil Soils 12:28-32

Crowley TJ (1991) Ice-Age Methane Variations. Nature 353:122-123

Dalton H (1992) Methane oxidation by methanotrophs: physiological and mechanistic implications. pp. 85-114. In, Murrell JC and Dalton H (eds.), Methane and methanol utilizers. Plenum Press, New York

Dawes EA (1985) Starvation, survival and energy reserves. pp. 43-79. In, Fletcher M and Floodgate GD (eds.), Bacteria in their natural environments. Academic Press, New York

Epp MA, Chanton JP (1993) Rhizospheric methane oxidation determined via the methyl fluoride inhibition technique. J Geophys Res 98(D):18413-18422

Fischer CR, Fisher JJ, Oremland RS, Bidigare RR (1987) The importance of methane in the metabolism of the bacterial symbionts of two deep-sea mussels. Mar Biol 96:59-71

Frenzel P, Thebrath B, Conrad R (1990). Oxidation of methane the oxic surface layer of a deep lake sediment (Lake Constance). FEMS Microbiol Ecol 73:149-158

Graham DW, Chaudhary JA, Hanson RS, Arnold RG (1993) Factors affecting competition between Type I and Type II methanotrophs in two-organism, continuous-flow reactors. Microb Ecol 25:1-17

Griffiths RP, Caldwell BA, Cline JD, Broich WA, Morita RY (1982) Field observations of methane concentrations and oxidation rates in the southeastern Bering Sea. Appl Environ Microbiol 44:435-446

Harrits SM, Hanson RS (1980) Stratification of aerobic methane-oxidizing organisms in Lake Mendota, Madison, Wisconsin Limnol Oceanogr 25:412-421

King, G.M. 1990a. Regulation by light of methane emission from a Danish wetland. Nature (London) 345:513-515

King GM (1990b) Dynamics and controls of methane oxidation in a Danish wetland sediment. FEMS Microbiol Ecol 74:309-323

King GM (1992) Ecological aspects of methane oxidation, a key determinant of global methane dynamics. Adv Microbial Ecol 12:431-468

King GM (1993) Ecophysiological characteristics of obligate methanotrophic bacteria and methane oxidation *in situ*. pp. 303-313. In, Murrell JC, Kelly DP (eds.), Microbial growth on C1 compounds. Intercept Scientific Publications, Andover

King GM (1994) Associations of methanotrophic bacteria with the roots and rhizomes of aquatic vegetation. Appl Environ Microbiol 60:3220-3227

King GM, Schnell S (1994) Enhanced ammonium inhibition of methane consumption in forest soils by increasing atmospheric methane. Nature 370:282-284

King GM, Skovgaard H, Roslev P (1990) Methane oxidation in sediments and peats of a sub-tropical wetland, the Florida Everglades. Appl Environ Microbiol 56:2902-2911

Koschorreck M, Conrad R (1993) Oxidation of atmospheric methane in soil: measurements in the field, in soil cores and in soil samples. Glob Biogeochem Cyc 7:109-121

Kuznetsov SI, Dubinina GA, Lapteva NA (1979) Biology of oligotrophic bacteria. Annu. Rev. Microbiol. 33:377-387.

Lees V, Owens NJP, Murrell JC (1991) Nitrogen metabolism in marine methanotrophs. Arch of Microbiol 157:60-65

Lidstrom ME (1988) Isolation and characterization of marine methanotrophs. Antonie van Leeuwenhoek 54:189-199

Moore TR, Knowles R (1989) The influence of water table levels on methane and carbon dioxide emissions from peatland soils. Can. J. Soil. Sci. 69:33-38

Morita RY (1985) Starvation and miniaturization of heterotrophs, with special emphasis on maintenance of the starved viable state. pp. 111-130. In, Fletcher M and Floodgate GD (eds.), Bacteria in their natural environments. Academic Press, New York

Mosier A, Schimel D, Valentine D, Bronson K, Parton W (1991) Methane and nitrous oxide fluxes in native, fertilized and cultivated grasslands. Nature (Lond) 350:330-332.

Poindexter JS (1981) Oligotrophy: fast and famine existence. Adv Microb Ecol 5:63-90

Reeburgh WS, Whalen SC, Alperin MJ (1993) The role of methylotrophy in the global methane budget. pp. 1-14. In, Murrell JC and Kelly DP (eds.), Microbial growth on C1 compounds. Intercept Ltd., Andover

Remsen CC, Minnich EC, Stephens RS, Buchholz L, Lidstrom ME (1989) Methane oxidation in Lake Superior sediments. J. Great Lakes Res. 15:141-146

Roslev P, King GM (1993) TTC, a new water-soluble formazan dye, and applications in starvation responses of bacteria. Appl Environ Microbiol 59:2891-2896

Roslev P, King GM (1994) Survival and recovery of methanotrophic bacteria under oxic and anoxic conditions. Appl Environ Microbiol 60:2602-2608

Roslev P, King GM (1995) Aerobic and anaerobic starvation metabolism in methanotrophic bacteria. Appl Environ Microbiol 61:1563-1570

Roulet NT, Ash R, Quinton W, Moore T (1993) Methane flux from drained northern peatlands: effect of a persistent water table lowering on flux. Global Biogeochem Cyc 7:749-769

Rudd JWM, Hamilton RD (1978) Methane cycling in a eutrophic shield lake and its effects on whole lake metabolism. Limnol Oceanogr 23:337-348

Schnell S, King GM (1994) Mechanistic analysis of ammonium inhibition of atmospheric methane consumption in forest soils. Appl Environ Microbiol 60:3514-3521

Schütz H, Schröder P, Rennenberg H (1991) Role of plants in regulating the methane flux to the atmosphere. pp. 29-64. In, Sharkey TD, Holland EA, Mooney HA (eds.), Trace gas emissions by plants. Academic Press, Inc., New York

Sieburth JM, Johnson PW, Eberhardt MA, Sieracki ME, Lidstrom M, Laux D (1987) The first methane-oxidizing bacterium from the upper mixing layer of the deep ocean, *Methylomonas pelagica* sp. nov. Curr Microbiol 14:285-293

Sinke AJC, Cottaar FHM, Buis K, Keizer P (1992) Methane oxidation by methanotrophs and its effects on the phosphate flux over the sediment-water interface in a eutrophic lake. Microb Ecol 24:259-269

Steele LP, Dlugokencky EJ, Lang PM, Tans PP, Martin RC, Masarie KA (1992) Slowing down of the global accumulation of atmospheric methane during the 1980s. Nature (Lond) 358:313-315

Steudler PA, Bowden RD, Mellilo JM, Aber JD (1989) Influence of nitrogen fertilization on methane uptake in temperate forest soils. Nature (London) 341:314-316

Street-Perrott FA (1992) Atmospheric methane: tropical wetland sources. Nature (Lond) 355:23-24

Striegl RG, McConnaughey TA, Thorstenson DC, Weeks EP, Woodward JC (1992) Consumption of atmospheric methane by desert soils. Nature (Lond) 357:145-147

Tate CM, Striegl RG (1993) Methane consumption and carbon dioxide emission in tallgrass prairie: effects of biomass burning and conversion to agriculture. Global Biogeochem Cyc 7:735-748

Ward BB, Kilpatrick KA (1990) Relationship between substrate concentration and oxidation of ammonium and methane in a stratified water column. Con Shelf Res 10:1193-1208

Whalen SC, Reeburgh WS (1990) Consumption of atmospheric methane by tundra soils. Nature (Lond) 346:160-162

Whittenbury R, Davies SL, Davey JF (1970a) Exospores and cysts formed by methane-utilizing bacteria. J Gen Microbiol 61:219-226

Whittenbury R, Phillips KC, Wilkinson JF (1970b) Enrichment, isolation and some properties of methane-utilizing bacteria. J Gen Microbiol 61:205-218

Williams ST (1985) Oligotrophy in soil: fact or fiction? pp. 81-110. In, Fletcher M and Floodgate GD (eds.), Bacteria in their natural environments. Academic Press, New York

Methane Production and Oxidation in Soils and Sediments

David B Nedwell
Department of Biology
University of Essex
Colchester CO4 3SQ UK

Introduction

Methane is the major end-product of anaerobic organic matter degradation in the biosphere. Its importance as a radiatively active trace gas in the atmosphere, increasing at about 1% y^{-1}, has focussed attention on human perturbation of the global CH_4 budget, and on the processes which give rise to or remove methane in different environments. Estimates of methane efflux to the atmosphere from the major biomes (IPCC, 1990; 1992; and see Table 1) show that on a global basis the largest source of methane emitted to the atmosphere is from the extensive wetlands which cover the subpolar regions in the Northern hemisphere. After this large natural source, anthropogenic inputs from the coal and petroleum industries, rice paddies and landfill sites all vastly exceed natural sources. These emissions are, however, net effluxes, representing the gross methane production minus any oxidation of methane which occurs before the methane is emitted to the atmosphere. Much of the methane that is formed anaerobically is reoxidized by methane-oxidizing (methanotrophic) bacteria before it is emitted to the atmosphere. Early estimates (e.g. Higgins et al, 1981) suggested that <1% of the biologically formed methane was emitted to the atmosphere. More recent calculations (Reeburgh et al, 1993) have attempted to correct net methane emissions by the best estimates of methane consumption in different environments to achieve a better estimate of global gross methane formation.

Of the major sources, a large part of the methane formed in wetlands and lakes and oceans is reoxidized before emission, but not that derived from the coal/petroleum industry. Overall, global gross methane production was

NATO ASI Series, Vol. I 39
Microbiology of Atmospheric Trace Gases
Edited by J. Colin Murrell and Donovan P. Kelly

calculated to be ~1188 $Tg.y^{-1}$, but emission to the atmosphere is only 500 $Tg.y^{-1}$. That is, about 60% of the methane formed globally is oxidized before it is emitted to the atmosphere. Much of the current research into the ecology of methanotrophic bacteria is directed towards trying to elucidate the factors which control this microbiological attenuation of the methane flux to the atmosphere, and to try to predict how environmental changes may perturb the balance between CH_4 production and oxidation in the biosphere.

Table 1. Net atmospheric CH_4 emission, global consumption and gross global production (Tg CH_4. y^{-1}). [after Reeburgh et al, 1993].

Source/sink	E + Net atmospheric budget	C = Global consumption	P Global production
Animals	80	0	80
Wetlands	115	27	142
Rice production	100	477	577
Biomass burning	55	0	55
Termites	20	24	44
Landfills	40	22	62
Oceans,freshwater	10	75.3	85.3
Hydrates	5 ?	5	10
Coal production	35	0	35
Gas production	40	18	40
Total Sources	500		
Chemical destruction		-450	
Soil consumption	-10	40	40
Total sinks	-460	688.3	
		Total production	1188.3

CH_4 may be oxidized microbiologically both aerobically or anaerobically. Anaerobic oxidation of methane was first reported (Reeburgh, 1976; Martens and Berner, 1977) at the base of the sulphate reduction zone in deep marine sediments. Here at great depth the slow residual activity of the sulphate-reducing bacteria (SRB) permitted the production of methane to occur, and peaks of sulphate reduction activity were reported in the depth horizons where the concentration gradients of methane and sulphate crossed over. The implication was that methane was acting as a substrate for SRB, and was being oxidized according to:

$$CH_4 + SO_4^{2-} \Rightarrow HS^- + HCO_3^- + H_2O$$

No bacteria have yet been isolated which are capable of carrying out this reaction, and it has been speculated that possibly a consortium of bacteria bring about the response.

In such deep marine sediments, where ebullition of CH_4 does not occur, the majority of the deep CH_4 is oxidized anaerobically (Reeburgh, 1976; Alperin and Reeburgh, 1984; Iversen and Jørgensen, 1985) and is never emitted to the atmosphere. However, seeps of natural gas may occur where greatly enhanced rates of anaerobic CH_4 oxidation are detectable, but where the fast rate of CH_4 transport in the gas phase still results in CH_4 output. Hydrocarbon-bearing sedimentary basins occupy 10-30% of the continental shelves, and it is suggested that they all have gas seepages (Hovland et al, 1993). Hovland et al (1993) calculated that 8-65 Tg $CH_4.y^{-1}$ passed through the seabed of the continental shelves to the water column. Hovland and Judd (1992) estimated the sources of natural gas seeps from shallow (< 250 m water depth) submarine seabed overlying hydrocarbon-rich areas as 145 Tg CH_4. y^{-1} (9.1 x 10^{12} moles $CH_4.y^{-1}$) : a figure as great as for emission from global wetlands (110 Tg CH_4. y^{-1} ; IPCC, 1992), for example. CH_4 oxidation and SO_4^{2-} reduction rates were very high immediately adjacent to a gas seep, and declined with increased distance from a seep. In control cores outside the seep area anaerobic CH_4 oxidation was negligible, and SO_4^{2-} reduction at the rates typical of normal intertidal sediments. Even with the extremely high rates of anaerobic CH_4 oxidation adjacent to seeps only about 11% of the CH_4 flux was anaerobically oxidized, the majority being emitted as gas bubbles to the atmosphere. The majority of CH_4 in these sediments was in the gas phase and by-passed biological oxidation.

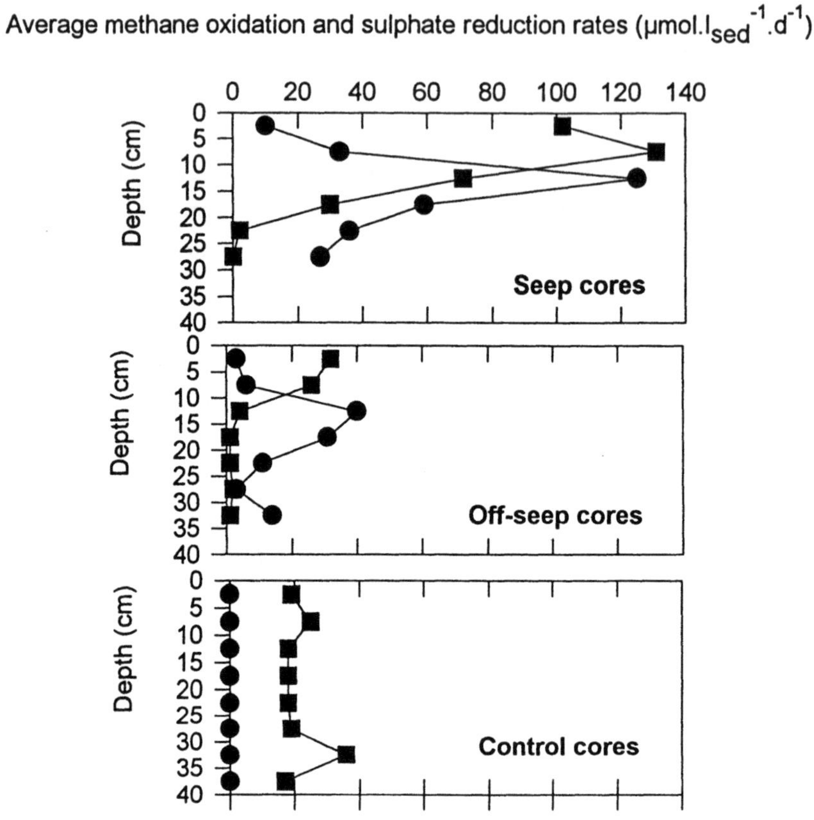

Figure 1. Rates of anaerobic methane oxidation (•) and sulphate reduction (■) in sediments of an intertidal gas seep area. Seep cores are immediately over seeps; off-seep cores adjacent to seeps; and control cores completely outside the gas seep area. (Schuster, Nedwell and Dando, unpublished data).

Aerobic CH_4 oxidation is mediated by a methanotrophic bacterial community which adapts to the presence or absence of CH_4. The ecological status of a microbial community can be described by measurements of its *in situ* activity, or potential activity, which reflect the community's capabilities. This may be independent of the phylogenetic structure of that community. A commonly used approach has been to describe a community kinetically, in terms of its maximum capacity (V_{max}) for a reaction, and its half-saturation constant (K_m) for the substrate of that reaction. A low K_m value has been held to indicate a high affinity for the substrate.

Response of the methanotroph community to the presence or absence of high concentrations of CH_4 has been described in a number of studies. Bender and Conrad (1992) showed the presence in fresh soils of a methanotroph community with high affinity (K_{CH4} 13-470 nM CH_4), but low capacity (V_{max} 2.1 to 150 nmol $CH_4.h^{-1}.g^{-1}$ dry wt soil): but in soils exposed to elevated CH_4 a low affinity (K_{CH4} 1740 to 27900 nM CH4), high capacity (V_{max} 270 to 3690 nmol $CH_4.h^{-1}.g^{-1}$ dry wt soil) methanotroph community developed. In specialized environments exposed to high CH_4 the aerobic methanotroph community may develop to extreme extent by selection of large numbers of obligate methanotrophs. In landfill cover soil (Kightley et al, 1995) a methanotrophic capacity as great as 10.4 moles $CH_4.m^{-2}.d^{-1}$ developed, with a K_{CH4} of 32 μM and V_{max} 998 nmol $CH_4.g^{-1}soil.h^{-1}$. The capacity for CH_4 oxidation is so great that a layer of cover soil >50cm deep above landfill seems to be sufficient to essentially prevent CH_4 efflux: an important factor when landfill is a major source of atmospheric CH_4. A consequence of this is that output of CH_4 from landfill sites may be significantly overestimated if it is calculated without allowing for CH_4 oxidation.

The kinetic characteristics of these high capacity communities are typical of those of the obligate methanotrophs available in laboratory pure cultures: not surprisingly, as these pure cultures have usually been isolated after exposure to elevated CH_4 concentrations. In pristine soil environments, with no elevated concentrations of CH_4 present, a methanotrophic community may still be present which has a limited capacity (small V_{max}) for CH_4 oxidation, but a high affinity (low K_{CH4}) for CH_4. The affinity for CH_4 may be such that the methanotroph community very effectively mops up any

CH_4 present, even at very low CH_4 concentrations. For example, methanotrophs in the surface 2 mm deep oxic layer of sediment in Lake Constance were able to remove >98% of the CH_4 fluxing upwards from deeper anoxic layers before it was emitted from the surface, indicating a very high affinity for CH_4 (Frenzel et al, 1990).

The use of molecular techniques has permitted better understanding of the phylogeny and, to some extent, the ecology of the obligate methanotrophs (Hanson et al, 1993; Green, 1993; Brusseau et al, 1994), but the phylogeny and ecology of the high affinity, low capacity methanotrophs remains obscure, although their ecological function is important. It might be questioned whether such microorganisms are obligate methanotrophs, or rather are facultative types which can persist in the absence, or at very low concentrations, of CH_4 by a more flexible, facultative nutritional strategy, where CH_4 is only one of a range of substrates that may be used to obtain the energy necessary for survival. Such a flexible, non-specialist strategy is typical of oligotrophic types surviving in nutrient-limited environments where mixotrophic growth on a variety of substrates optimises the sequestration from the environment of sufficient energy to permit survival (Poindexter, 1981, 1987; Upton and Nedwell, 1989). Low specificity of soluble *methane monooxygenase (MMO)*, the key enzyme in CH_4 oxidation, is known to permit some use of other substrates such as halomethanes, halogenated hydrocarbons and trichloroethylene, but may also sequester other organic molecules present in natural environments. Isolation from landfill cover soil of methanotrophs on silica gel agar with no organic matter present (Jones and Nedwell, 1993) revealed that most of the isolates were apparently not obligate methanotrophs, but facultative types subsequently shown to be capable of using organic substrates also. We do not even know for sure that any facultative methanotrophs are bacteria. There are a number of unsubstantiated reports of yeasts (Wolf and Hanson, 1979; Tani, 1984) and fungi (Adamse et al, 1972) as well as actinomycetes (Reed and Dugan, 1987) which were reported to use CH_4. The extent to which molecular probes, developed for and tested against the obligate methanotrophs, might cross-react with and detect other, possible phylogenetically dissimilar, facultative methanotrophs remains open to question.

Our incomplete understanding of the phylogenetic range and physiological capabilities of CH_4-utilizing microorganisms is also well illustrated by the work of King and coworkers (King et al, 1990) who have shown that the methanotrophic community in an Everglade peat soil contained a component that rapidly disappeared under anoxic conditions; but a second part of the community survived anoxia for extended periods of time and rapidly reestablished CH_4 oxidation when oxic conditions returned. More recent work (Roslev and King, 1994, 1995) suggests that in the absence of CH_4 obligate methanotrophs can survive long periods under anaerobic conditions, although they survive less well aerobically. Aerobic methanotrophic potential survived in soils purged with N_2 for >90 days in the total absence of CH_4 (Kightley and Nedwell, unpublished data) Under anaerobic conditions in the absence of CH_4 methanotrophs apparently enter a dormant state where endogenous metabolism is severely reduced, and energy is provided by fermentation of intracellular organic molecules (Roslev and King, 1995).

The removal of CH_4 by oxic methanotrophs in peat can be measured by headspace removal of CH_4, which can be shown to conform to saturation kinetics, with a K_{app} for CH_4 of 14 to 44 μM, depending upon the site. Variation of the V_{max} for aerobic CH_4 oxidation (the aerobic CH_4 oxidation potential) with depth in the peat can be used to show that there appears to be an apparent peak in the V_{max} at about 5-6 cm below the water table, which is surprising for an oxic process when the water table essentially represents the oxic/anoxic boundary (Figure 2). How can this paradox be resolved?
While the ability of methanotroph communities to remove CH_4 has been related to the ambient CH_4 concentration, the oxidation of CH_4 is a bisubstrate reaction depending upon the concentration of both CH_4 and O_2, according to the equation

$$\text{Rate of } CH_4 \text{ oxidation} = V_{max}[C_{CH_4}/C_{CH_4}+K_{CH_4}][C_{O_2}/C_{O_2}+K_{O_2}]$$

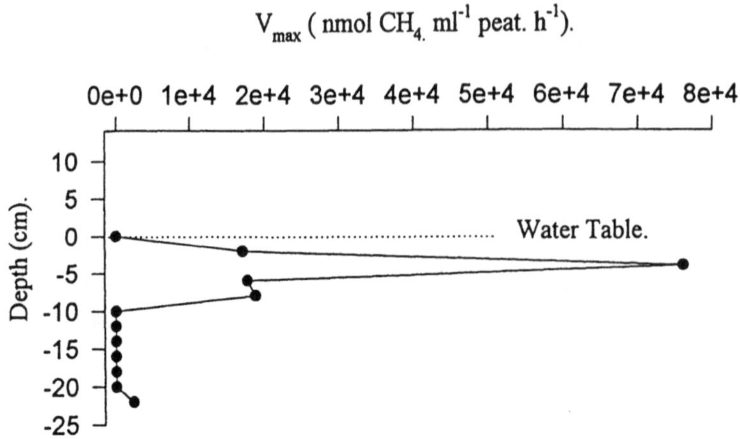

Figure 2. Distributions of aerobic methane oxidation potential (V_{max}) with depth in peat. The water table represents the oxic/anoxic interface (from Nedwell and Watson, in press).

The implication of this equation is that Liebig's Law of the Limiting Factor does not hold true, and methane oxidation rate may be simultaneously regulated by both CH_4 and O_2 if both substrates are at low concentration. As CH_4 oxidation occurs typically at the interface between oxic and anoxic layers, where both CH_4 and O_2 are at low concentrations, the affinity of the methanotrophs for both substrates may be important in regulating the rate of CH_4 oxidation which occurs. Figure 3 illustrates the effect of varying O_2 concentrations on O_2 removal by bacteria in peat, in both the absence and presence of CH_4. Saturation kinetics was observed for O_2 removal, with a K_{app} for O_2 of 33μM, and respiration becoming O_2-limited at concentrations ≈250μM O_2. About 80% of the O_2 uptake in the peat samples was driven by CH_4 oxidation when excess CH_4 was present, the balance by aerobic respiration of organic substrates from the peat. The difference between the two plots gives the kinetics for CH_4 oxidation alone.

The kinetic parameters measured in this work were used in a reaction diffusion model of a peat system to examine the factors controlling the distribution of gases and activity in these systems. The modelling has been carried out by Jon Arah (Institute for Terrestrial Ecology, Edinburgh) and Karl Steven (Scottish Agricultural College, Edinburgh) under the TIGER programme of the Natural Environment Research Council, U.K. The model includes terms for O_2 respiration of organic matter, CH_4 production and oxidation , and the effects of temperature upon these reactions.

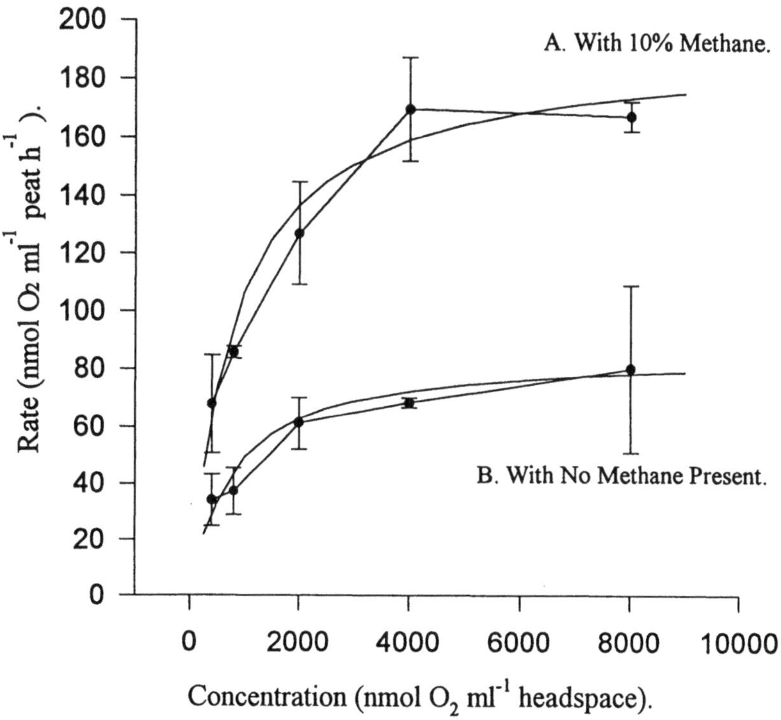

Figure 3. Kinetics of O_2 removal by peat in the presence (●) and absence (■) of methane. Curves show best fit of data by Eadie-Hofstee plots.

For O_2 distribution

$$\frac{d}{dx}D1(x)\frac{dO}{dx}-R(O)-Q(O,C)=0$$

For CH_4 distribution

$$\frac{d}{dx}D2(x)\frac{dC}{dx}-2Q(O,C)+I(O)P(x)Ap(T)=0$$

where O and C are the O_2 and CH_4 concentrations respectively; D1(x) and D2(x) are diffusion coefficients of O_2 and CH_4 in water; P(x)Ap(T) is the CH_4 production rate combined with an Arrhenius function for methanogenesis (150 $kJ.mole^{-1}$); and I(O) is an inhibition function to describe the inhibitory effect of O_2 upon CH_4 formation.

$$I(O)=0.5*(1+\cos(O*pi/Oth)$$
$$for O<2.0th, O>=2.0th$$

The cosine nature of the inhibition function reflects the fact that it is an areal average i.e. at some points O_2 >threshold and no CH_4 production, while at other points O_2 <threshold and CH_4 produced.

The aerobic respiration is described by

$$R(O)=Ar(T).R_{max}(x)\frac{O}{(O+rKM)}$$

where R_{max} is maximum O_2 uptake rate; rKM is the half saturation constant for O_2; Ar(T) is the Arrhenius function for respiration (50 $kJ.mole^{-1}$).

CH_4 oxidation by

$$Q(O,C)=Aq(T).Q_{max}(x)\ \frac{O \qquad C}{(0+qKMo)\ (C+qKMc)}$$

where Q_{max} is maximum CH_4 oxidation potential; qKMo and qKMc are half saturation constants for O_2 and CH_4 respectively; Aq(T) is an Arrhenius function for CH_4 oxidation (50 $kJ.mole^{-1}$).

CH_4 formation down the peat profile is described subjectively from measured rates of CH_4 production, and an Arrhenius function (150 $kJ.mole^{-1}$) to describe the effect of temperature. This is a large temperature effect, but is that measured in peat. In addition, gas transport into (O_2) and out of (CH_4) the peat via roots can be included. Root distribution is described horizontally by a Gaussian distribution, with a peak at 8 cm depth. Figure 4 shows the steady state profiles of gases and microbial activities which are predicted within the peat in a model run at 5°C, with and without the presence of gas transport by the roots. The main features to be noted are that in the absence of roots CH_4 oxidation is focussed near the peat surface, at the interface between the concentration gradients of O_2 and CH_4, intercepting the vertically diffusing CH_4. When root transport of gases is included the CH_4 profile does not come so close to the peat surface, and the concentration profiles of O_2 and CH_4 do not overlap. More importantly, the transport of O_2 through the roots supports considerable aerobic CH_4 oxidation in the root zone despite the fact that elevated O_2 concentrations are not predicted in the bulk soil at this depth. The production of CH_4 is, however, considerably reduced by low concentrations of O_2 below 10cm, which are sufficient to inhibit CH_4 formation. The important point, though, is that the peak of aerobic CH_4 oxidation potential that we measured at 6 cm depth in the peat, in the apparently anoxic zone, is not an artifact but may be explained by the transport of O_2 through plant roots sustaining aerobic CH_4 oxidation in the aerobic rhizosphere around the roots. Such root transport of O_2 can have a profound effect, both by inhibiting the formation of CH_4, by stimulating CH_4 oxidation, and by facilitating vertical transport of gases. The model predicts that CH_4 emission from the peat will be 0.4 $\mu moles.m^{-2}.h^{-1}$ without roots, and 33 $\mu moles.m^{-2}.h^{-1}$ when roots are included in the model.

The soil sink for CH_4 may, however, be disturbed by a variety of factors. It has been demonstrated that undisturbed forest, grassland and tundra soils all have a capacity to

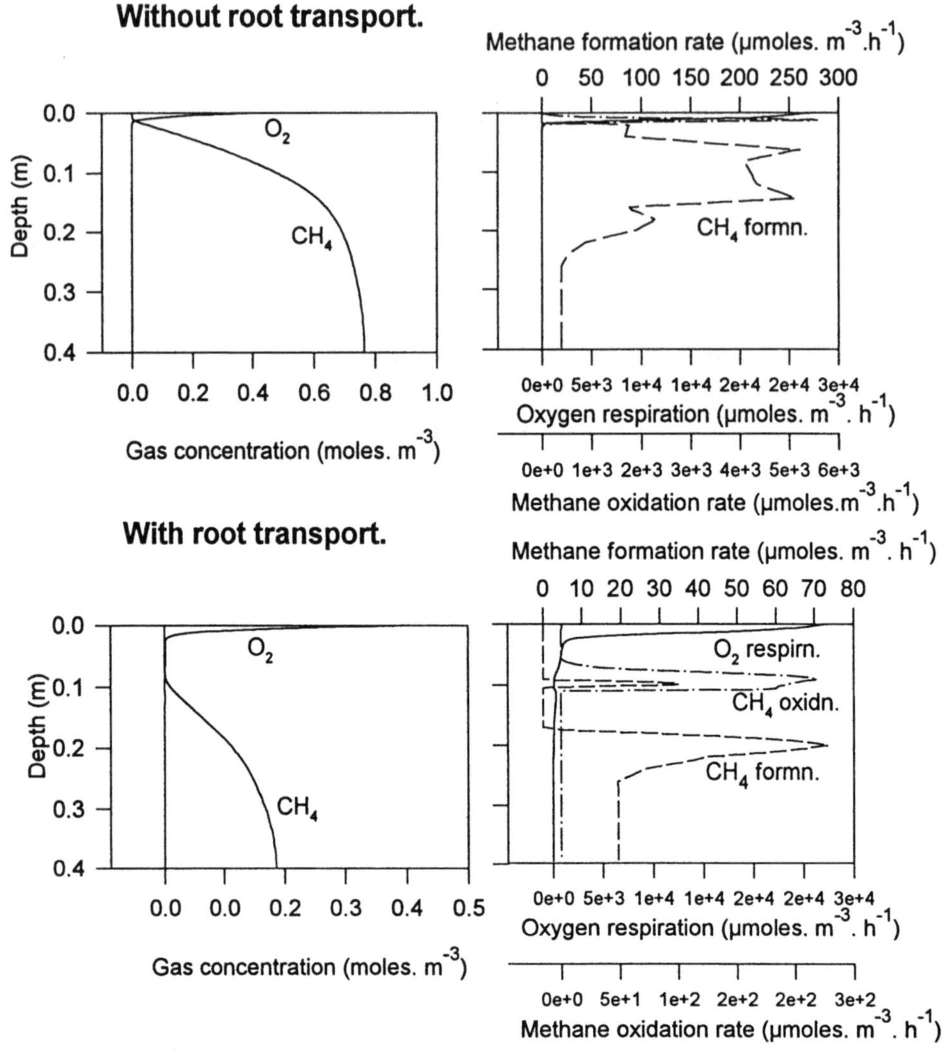

Figure 4. The vertical concentration profiles of gases and of microbial activities in peat at 5°C predicted by a reaction/diffusion model with and without gaseous transport in roots.

remove CH_4 from the overlying air (Steudler et al, 1989; Whalen and Reeburgh, 1990; Mosier et al, 1991). This capacity is the result of the activity of methanotrophs, with a high affinity for CH_4, oxidizing CH_4 within surface soil and acting as a sink for CH_4 from the overlying air. This soil CH_4 sink (15 - 45 Tg $CH_4.y^{-1}$) is the largest biological sink of atmospheric CH_4. However, it has also been demonstrated (e.g. Steudler et al, 1989; Mosier et al, 1991) that cultivation and fertilization of soils results in inhibition of CH_4 oxidation, and this important CH_4 sink is lost. This effect has been attributed to the presence in soils of NH_4^+ (e.g. King and Schnell, 1994), derived from fertilizer addition, which inhibits *MMO* (Carlsen et al, 1991). Ammonium acts as a competitive inhibitor of CH_4 oxidation, although the mechanism of NH_4^+ inhibition for soluble *MMO* is more complicated (Carlsen et al, 1991). There is little doubt that ammonium addition does depress the capacity of soil methanotroph microflora to oxidize CH_4. The long term experiments at Rothamsted show that in soil subjected to addition of NH_4^+ over 110 years there is virtually no methanotrophic activity, whereas addition of sodium nitrate over the same period has no such inhibitory effect (Hütsch et al, 1993, 1994; Willison et al, 1995). However, the degree of inhibition of methanotrophy in these soils was also affected by other factors such as soil pH, and the net effect of fertilization upon CH_4 oxidation in soil is not clear.

However, this specific inhibition of *MMO* by NH_4^+ appears to be only one mechanism affecting CH_4 oxidation in soils. A number of studies have now shown that a proportion of the inhibition of CH_4 oxidation in fertilized soil is sometimes not attributable to the presence of NH_4^+ in the soil, but to some other effect associated with fertilizer application. For example (Figure 5), in landfill cover soil, where CH_4 oxidation was active, the addition of NH_4Cl solution inhibited CH_4 oxidation, but so did addition of KCl or KNO_3. Rather than specific inhibition by NH_4^+, there seemed to be a non-specific inhibition of CH_4 oxidation by a generally increased ionic concentration. Nesbit and Breitenbeck (1992) showed that addition of 7 μmol $NH_4Cl.g^{-1}$soil inhibited CH_4 oxidation, and continued even after NH_4 had disappeared from the soil, but addition of KNO_3 or KCl also inhibited CH_4 oxidation by 70-75% of the unamended controls. They attributed this to susceptibility of methanotrophs to 'increased osmotic potential'. Crill et al (1994) similarly reported non-specific inhibition of CH_4 oxidation by addition of fertilizer to peat soil, and 80% of the inhibition reported by Adamsen and King (1993) was apparently not due to NH_4^+ but to a nonspecific effect. Elucidation of the net effect of fertilizer application is complicated by the fact that nitrifiers are also able to oxidize CH_4 (Jones and Morita, 1983). Selection of a larger nitrifying community in soil by addition of NH_4^+ indeed may inhibit CH_4 oxidation by methanotrophs, but there may be some degree of compensation by increased CH_4 oxidation by nitrifiers. Further research is required to finally elucidate the various factors which contribute to the loss of the CH_4 sink in cultivated soils.

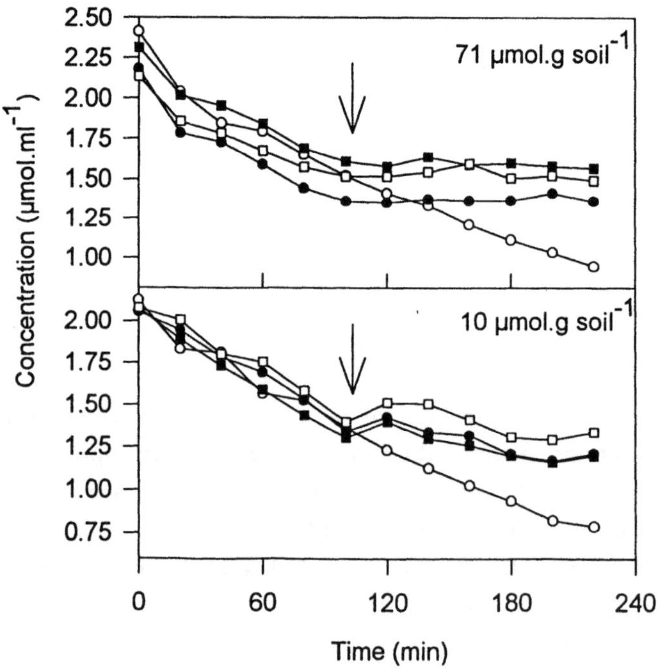

Figure 5. Effect on CH_4 oxidation of addition of distilled water (○), ammonium chloride solution (■), potassium nitrate solution (•) or potassium chloride solution (□) to landfill cover soil.

References

Adamse AD, Hoeks J, DeBont JAM, van Kesel JF (1972) Microbial activities in soils near natural gas leaks. Arch fur Mikrobiol 83:31-35

Adamsen APS, King GM (1993) Methane consumption in temperate and subarctic forest soils: rates, vertical zonation and responses to water and nitrogen. Appl Environ Microbiol 59:485-490

Alperin MC, Reeburgh WS (1984) Geochemical observations supporting anaerobic methane oxidation. In, *Microbial growth on C_1 compounds* (eds RL Crawford and RS Hanson) American Society for Microbiology Washington D.C

Alperin MC, Reeburgh WS (1985) Inhibition experiments on anaerobic methane oxidation. Appl Environ Microbiol 50:940-945

Bender M, Conrad RC (1992) Kinetics of CH_4 oxidation in oxic soils exposed to ambient air or high CH_4 mixing ratios. FEMS Microbiol Ecol 101:261-270

Brusseau GA, Bulygina ES, Hanson RS (1994) Phylogenetic analysis and development of probes for differentiating methylotrophic bacteria. Appl Environ Microbiol 60:626-636

Carlsen HN, Joergensen L, Degn H (1991) Inhibition by ammonium of methane utilization in *Methylococcus capsulatus* (Bath). Appl Microbiol Biotechnol 35:124-127

Crile PM, Marti Kainen PJ, Nykanen H, Silvola J (1994) Temperature and N fertilization effects on methane oxidation in a drained peatland soil. Soil Biol Biochem 26:1331-1339

Frenzel P, Thebrath B, Conrad R (1990) Oxidation of methane in the oxic surface layer of a deep lake sediment (Lake Constance). FEMS Microbiol Ecol 73:149-158

Green PN (1993) Overview of the current state of methylotroph taxonomy. In, Microbial growth on C_1 compounds (eds. JC Murrell and DP Kelly) 253-265. Intercept Ltd., Andover, UK

Hanson RS, Bratina BJ, Brusseau GA (1993) Phylogeny and ecology of methylotrophic bacteria. In, Microbial growth on C_1 Compounds (eds. JC Murrell and DP Kelly), 285-302. Intercept Ltd., Andover, UK

Higgins IJ, Best DJ, Hammond RC, Scott D (1981) Methane-oxidizing microorganisms. Microbiol Rev 45:556-590

Hovland M, Judd AG (1992) The global production of methane from shallow submarine sources. Continental Shelf Res 12:1231-1238

Hovland M, Judd AG, Burke RA Jr (1993) The global flux of methane from shallow submarine sediments. Chemosphere 26:559-578

Hütsch BW, Webster CP, Powlson DS (1993) Long term effects of nitrogen fertilization on methane oxidation in soil of the Broadbalk wheat experiment. Soil Biol Biochem 10:1307-1315

Hütsch BW, Webster CP, Powlson DS (1994) Methane oxidation in soil as affected by land use, soil pH and N fertilization. Soil Biol Biochem 12:1613-1622

Intergovernmental Panel on Climate Change, 1990. Climate change (eds. JT Houghton, GJ Jenkins and JJ Ephraums) Cambridge University Press

Intergovernmental Panel on Climate Change, 1992. Climate change 1992. (eds. JT Houghton, BA Callender and SK Varney) Cambridge University Press.

Iversen N, Jørgensen BB (1985) Anaerobic methane oxidation rates at the sulfate-methane transition in marine sediments from the Kattegat and Skaggerrak (Denmark). Limnol Oceanogr 30:944-955

Jones RD, Morita RY (1983) Methane oxidation by *Nitrosococcus oceanus* and *Nitrosomonas europaea*. Appl Environ Microbiol 45:401-410

Jones HA, Nedwell DB (1993) Methane emission and methane oxidation in landfill cover soil. FEMS Microbiol Ecol 102:185-195

Kightley D, Nedwell DB, Cooper M (1995) Capacity for methane oxidation in landfill cover soils measured in laboratory-scale soil microcosms. Appl Environ Microbiol 61:592-601

King GM, Schnell S (1994) Effect of increasing atmospheric methane concentration on ammonium inhibition of soil methane consumption. Nature 370:282-284

King GM, Roslev P, Skovgaard H (1990) Distribution and rate of methane oxidation in sediments of the Florida Everglades. Appl Environ Microbiol 56:2902-2911

Martens CS, Berner RA (1977) Interstitial water chemistry of anoxic Long Island Sound sediments I. Dissolved gases. Limnol Oceanogr 22:10-25

Mosier A, Schimel D, Valentine D, Bronson K, Parton W (1991) Methane and nitrous oxide in native, fertilized and cultivated grasslands. Nature 350:330-332

Nesbit SP, Breitenbeek GA (1992) A laboratory study of factors influencing methane uptake by soils. Agric Ecosyst Environ 41:39-54

Poindexter JS (1981) Oligotrophy: fast and famine existence. Adv Microb Ecol 5:63-89

Poindexter JS (1987) Bacterial responses to nutrient limitation. In, *Ecology of microbial communities* (eds. Fletcher M, Gray TRGG and Jones JG) 283-317, Cambridge University Press

Reeburgh WS (1976) Methane consumption in Cariaco Trench waters and sediments. Earth Planet Sci Lett 28:337-349

Reeburgh WS, Whalen SC, Alperin MJ (1993) The role of methylotrophy in the global methane budget. In, *Microbial Growth on C_1 compounds*. (eds JC Murrell and DP Kelly) 1-14, Intercept Ltd, Andover, UK

Reed WM, Dugan PR (1987) Isolation and characterization of the facultative methylotroph *Mycobacterium* ID-Y. J Gen Microbiol 133:1389-1395

Roslev P, King GM (1994) Survival and recovery of methanotrophic bacteria starved under oxic and anoxic conditions. Appl Environ Microbiol 60:2602-2608

Roslev P, King GM (1995) Aerobic and anaerobic starvation metabolism in methanotrophic bacteria. Appl Environ Microbiol 61:1563-1570

Steudler PA, Bowden RD, Melillo JM, Aber JD (1989) Influence of nitrogen fertlization on methane uptake in temperate forest soil. Nature 341:314-316

Tani Y (1984) Microbiology and biochemistry of methylotrophic yeasts. In, *Methylotrophs: microbiology, biochemistry and genetics* (ed CT Hou). CRC Press, Boca Raton, Florida

Upton, AC, Nedwell DB, (1989) Nutritional flexibility of oligotrophic and copiotrophic Antarctic bacteria with respect to organic substrates. FEMS Microbiol Ecol, 62:1-6

Whalen SC, Reeburgh WC (1990) Consumption of atmospheric methane by tundra soils. Nature 346:160-162

Willison TW, Webster CP, Goulding KWT, Powlson DS (1995) Methane oxidation in temperate soils: effects of land use and the chemical form of nitrogen fertilizer. Chemosphere 30:539-546

Wolf HJ, Hanson RS (1979) Isolation and characterization of methane-utilizing yeasts. J Gen Microbiol, 114:187-194

Methane Oxidation in Coastal Marine Environments

Niels Iversen
Environmental Engineering Laboratory
Aalborg University
Sohngaardsholmsvej 57
9000 Aalborg, Denmark

Methane oxidation has been observed in both the water column and sediments of marine environments. The number of publications is relatively limited but does indicate that methane oxidation is an important process regulating methane emission from the marine areas.

Table 1. Global methane budget with annual emissions for identified sources (Cicerone and Oremland. 1988)

	10^{12} g	
Emission source	**Annual release**	**Range**
Oceans	10	5-20
Natural wetlands	115	100-200
Rice paddies	100	60-170
Freshwaters	5	1-25
Enteric fermentation	80	65-100
Termites	40	10-100
Biomass Burning	55	50-100
Landfills	40	30-70
Coal mining	35	25-45
Gas drilling, venting, transmission	45	25-50
Methane hydrate destabilization	5	0-100
Total	540	400-640

The current global methane budgets estimate that the oceans contribute 10 Tg y^{-1}, and thus account for less than 2% of the global methane emission (Table 1). The emission for the oceans is largely based on Ehhalt's (1974) estimate of 4.7 to 20.7 Tg y^{-1} (Table 2). In a recent estimate by Lambert and Schmidt (1993), they calculated a methane emission of 10 Tg y^{-1} for

NATO ASI Series, Vol. I 39
Microbiology of Atmospheric Trace Gases
Edited by J. Colin Murrell and Donovan P. Kelly

the oceans using a methane concentration of 2.5 ± 0.8 nmol l^{-1} (STP) for the open oceans and 8.9 nmol l^{-1} (STP) for the coastal surface water.

The reported methane concentration in the coastal waters varies from the above mentioned oceanic concentrations and up to 1600 nmol l^{-1} (Lambert and Schmith, 1993; Sansone and Martens, 1978), which indicates that a large variation in emission rates can be expected. Taking natural methane seeps, and the known coastal areas with a high rate of methane release into account, increased the methane emission estimate to 50-60 Tg y^{-1} (Table 2)

Table 2. Estimates of methane emission from the oceans.

10^{12} g			
Total	**Coastal**	**Open ocean**	
4.7-20.7	0.7-14	4-6.7	Ehhalth (1974)
50-60	47-56	3-4	Lambert and Smith (1993)
	8-65		Hovland et al.(1993)

Hovland et al. (1993) estimated that the coastal zone could contribute as much as 65 Tg y^{-1}, if they included sediments with a high organic loading and natural methane seeps into the budget.

According to the estimates in table 2, the coastal zone clearly represents the major methane source in the marine environment. This is also what should be expected, as the coastal zone represents an area where high carbon input is expected both from *in-situ* primary production and from terrestrial sources.

Methane in the marine environment is oxidized by both aerobic and anaerobic methane oxidation. The aerobic methane oxidation is restricted to the water column and the narrow oxic zone at the sediment surface. Anaerobic methane oxidation is primarily found at the sulfate reduction zone in the sediment, but is also found in anoxic water columns.

Anaerobic methane oxidation

Anaerobic methane oxidation has been observed throughout the sulfate reduction zone in several marine sediments (Reeburgh. 1980; Devol, 1983; Iversen and Jørgensen, 1985; Alperin and Reeburgh, 1985). The rates of anaerobic methane oxidation are low (0.001 - 0.5 nmol cm^{-3} d^{-1}) at most of the sulfate reduction zone (Fig 1.) but show a large increase at the sulfate-methane transition zone, where a maximum of both sulfate reduction and methane oxidation (0.5 - 6 nmol cm^{-3} d^{-1}) is observed (Fig.1).

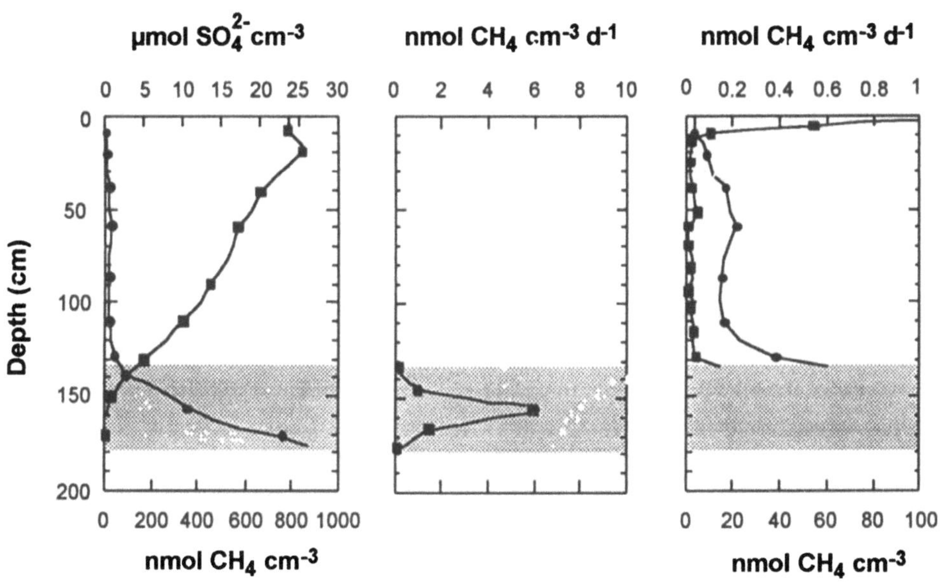

Fig. 1. Sulfate (B) and methane concentration profiles (J). Methane oxidation rates (B) measured at the sulfate transition zone and expansion of the methane concentration and methane oxidation profiles at the sulfate reduction zone. The shaded area show the location and extension of the sulfate-methane transition zone.(Redrawn from Iversen and Jørgensen, 1985).

The biological nature of the process has been confirmed (Iversen et al.,1987; Alperin and Reeburgh,1985), but the mechanism behind anaerobic methane oxidation still needs to be demonstrated.

No anaerobic methane oxidizers have yet been identified and it remains uncertain which electron acceptor is used for anaerobic methane oxidation. Sulfate has been suggested as the electron acceptor for anaerobic methane oxidation, mainly because coinciding peaks of anaerobic methane oxidation and sulfate reduction are normally observed at the sulfate-methane transition zone, where the highest activity of methane oxidation is found (Reeburgh. 1980; Devol, 1983; Iversen and Jørgensen, 1985; Alperin and Reeburgh, 1985).

Table 3. Anaerobic methane oxidation rates reported from different marine sediments.

	$\mu mol\ l^{-1}\ d^{-1}$		
	Sulfate reduction zone	**Sulfate - methane transition zone.**	Reference
Danish coastal	0-0.5	0.01-20	Iversen and Jørgensen (1985)
Scan Bay		0-8.2	Alperin and Reeburgh (1985)
Saanich Inlet		0.5-34	Devol (1983)
Black Sea		0-0.0026	Reburgh et al. (1992)

Specific inhibitors for sulfate reduction (molybdate) and methane production (bromo-ethanesulfonic acid) have failed to resolve the problem, as none of these inhibitors could terminate anaerobic methane oxidation (Alperin and Reeburgh,1985; Hoehler et al. 1994). These results indicate that anaerobic methane oxidation can be uncoupled from both sulfate reduction and methane production. In a recently published paper by Hoehler at al. (1994), it was suggested that the observed anaerobic methane oxidation rates are mediated by a consortium of methane producers and sulfate reducers. The methane producers are hypothesized to oxidize CH_4 to CO_2 and hydrogen. To make the process thermodynamically favorable, the hydrogen produced has to be removed which could be done by sulfate reducers. The reoxidation of methane by active methane producers has also been shown by Zehnder and Brock (1980), but the methane oxidation rate was only a very minor fraction of the methane production rate. Therefore, this process alone would not explain the observed net methane oxidation in marine sediments. Furthermore, a net methane oxidation can only be measured if there is active sulfate reduction in the sediment (Alperin and Reeburgh, 1985; Hoehler et al. 1994; Hansen et al. 1995).

The lack of inhibition of anaerobic methane oxidation when using known inhibitors for methane producers indicates that the methane producer participating in this consortium must behave differently than the known methane oxidizers, or use an enzymatic pathway for the reoxidation of methane that is not affected by these inhibitors. However, a bacterial consortium seems a reasonable explanation for the anaerobic methane oxidation activity measured in marine sediments, but the results could indicate that some bacteria other than the methane producers may be involved in the consortium.

Table.4. Integrated rates of methane oxidation calculated from the published methane oxidation rate profiles, and methane flux calculated from the published methane concentration profiles using a diffusion coefficient of 0.54 10^{-5} cm^2 s^{-1} (Iversen and Jørgensen, 1993)

	m mol m^{-2} d^{-1}		
	CH_4 oxidation	CH_4 flux	Reference
Danish coastal	0.8 - 1.2	0.2 - 0.7	Iversen and Jørgensen, 1985
Scan Bay	1.1	0.8	Alperin and Reeburgh, 1985
Saanich Inlet	1.8	1.9	Devol , 1983

Irrespective of the type of organism that is actually oxidizing methane anaerobically, the role of anaerobic methane oxidation as an efficient biological sink is well documented. Several studies have shown that a nearly complete oxidation of the upward methane flux takes place at the sulfate-methane transition zone (Reeburgh. 1980; Devol, 1983; Iversen and Jørgensen, 1984; Alperin and Reeburgh, 1985). The reported rates of anaerobic methane oxidation at the sulfate-methane transition zone are surprisingly uniform (Table 3). The integrated rate of anaerobic methane oxidation from the different sediments at the sulfate-methane transition also fall into a narrow range from 0.8 - 1.8 $m^{-2}d^{-1}$ (Table 4). This suggests that anaerobic methane oxidation is regulated by the same factors in the different sediments. What factors however, that regulate anaerobic methane oxidation in marine sediments can probably best be determined, when the process is clarified.

In marine sediments the sulfate-methane transition zone is typically located deep in the sediment, and low methane concentrations (1-10 nmol cm^{-3}) are normally found at the depths above the transition zone. At the depths where the sulfate concentration starts to be depleted, a rapid increase in methane concentration is observed and the methane concentration often exceeds saturation and causes bubble formation in the sediment. A methane profile like this can only exist if there is a net oxidation of the upward methane flux in the sediment (Martens and Berner, 1977). Thus, in a typical coastal marine sediment, anaerobic methane oxidation removes all the upward methane flux. However, in some sediments, part of the upward methane flux escapes the sulfate methane transition zone and increased methane concentrations (50-800 nmol cm^{-3}) can be measured at the sulfate containing depths (Fig. 2A).

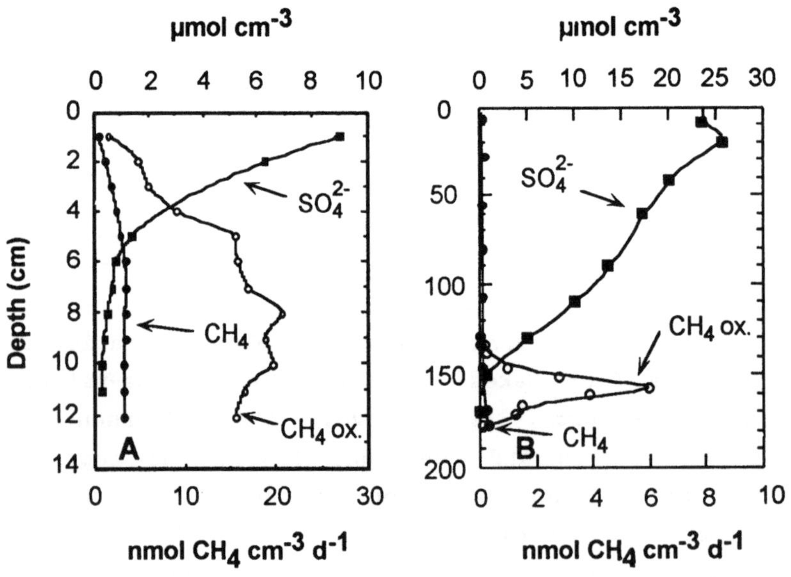

Fig. 2. Depth profiles of methane and sulfate concentrations and anaerobic methane oxidation rates from two stations with different sedimentation rates. A - Limfjorden, Denmark, more than 1 cm y^{-1}. B - Kattegat, Denmark, 0.16 cm y^{-1} (Redrawn from Iversen and Jørgensen, 1985).

There seems to be a relation between the sedimentation rate and the amount of methane that escapes oxidation in the sediment, probably because a higher sedimentation equals a higher carbon input. The depth at which sulfate depletion occurs moves upward as a result of an increased sedimentation rate is shown in Fig. 2.

In the more extreme situations, sulfate is depleted in the upper 10 cm. Methane bubbles can be observed immediately below this depth and a large amount of the methane is released as bubbles thereby escaping oxidation (Martens and Klump, 1980; Crill and Martens, 1983, Fenchel et al., 1995). It is not possible to say how high the carbon input to the sediment has to be in order to trigger methane bubble release. The few data from Danish coastal sediments shown in Fig. 3 indicate that the sediment may change from a diffusion controlled system to a bubble controlled system at a sedimentation rate around 1 cm y^{-1}.

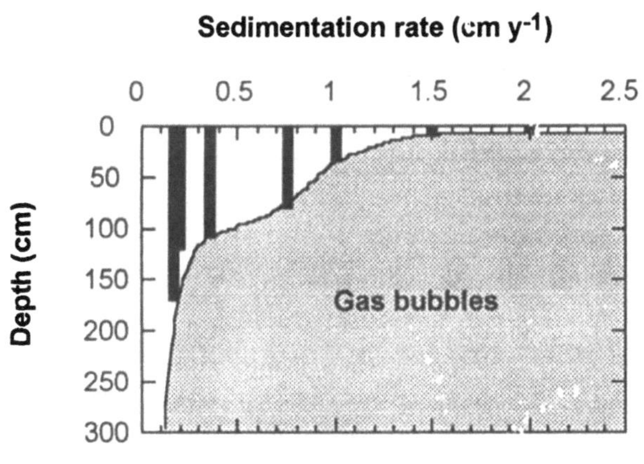

Fig. 3. Extension of the sulfate reduction zone at different sedimentation-rates. The filled bars represent the depths, where sulfate is present. The shaded area illustrates gas bubbles in the sediment at (STP). All results are from Danish Coastal sediments.

This has a marked implication for the amount of methane that is released to the water column. In very organic rich sediments, up to 50% of the methane can be released as bubbles

(Chanton et al. 1989). The importance of anaerobic methane oxidation in these organic rich sediments is difficult to evaluate because the methane profile in these types of sediments is not a direct reflection of the methane production. Comparing the calculated flux to the measured methane oxidation will, however, give a maximum estimate of the amount of methane oxidized. The two examples from organic rich sediments shown in table 5 suggest that less than 50% of the upward methane flux is oxidized. At both stations, bubbles are often seen breaking the water surface (Nordjyllands County, pers. comm), and a methane emission rate of 4.5 mmol $m^{-2}d^{-1}$ has been measured at the water surface (Fenchel et al. 1995). However, both stations are from fjords with anoxic bottom water and do not represent examples of the more open coastal zone.

Table 5. Estimates of methane oxidation and methane flux from two organic rich sediments.

	cm	m mol m^{-2} d^{-1}			
	Sulfate zone	CH_4 oxidation	Methane flux	% oxidized	Reference
Mariager Fjord	11	0.04	0.3-1.5	3 - 13	Fenchel et. al 1995
Virksund	12	0.6	1.4	43	This study

The upward movement of the sulfate methane transition zone could also be expected along salinity gradients, where the lower salinity will reflect a lower sulfate concentration in the water column. If this is correct, then estuaries could also be sites, where anaerobic methane oxidation has lost its importance, resulting in high methane emission rates.

Aerobic methane oxidation

Aerobic methane oxidation has been identified as a very important methane sink in freshwater sediments, where up to 99% of the sediment water methane flux is removed by aerobic methane oxidation at the sediment surface (King, 1990). Much to my surprise, aerobic methane oxidation rates have not been reported from any marine sediments except for one shallow estuary where *in-situ* methane oxidation rate measurements indicated that aerobic methane oxidation took place in the upper one cm of the sediment (Iversen and Blackburn, 1981).

Direct measurements of aerobic methane oxidation in several near shore marine sediments in Denmark showed that aerobic methane oxidation occurs at the sediment surface. The measurements were made in two sediment types, one type that represents organic poor sediment with a low methane concentration (less than 10 nmol cm^{-3}) and the other type represents a more organic rich sediment, where the methane concentrations in the upper 5 cm were high (up to 100 nmol cm^{-3}). The oxidation rates measured at both sediment types showed a very large variation (Table 6), but in general, the highest rates were observed at the stations with the highest methane concentration.

Table 6. Methane oxidation rates measured at two different sediment types in near shore marine sediments. Compared to rates reported from fresh-water sediments.

	μmol m^{-2} d^{-1}		
	Oxidation rates	**% oxidation**	Reference
Organic poor	0-180	0-99	This study
Organic rich	0-1100	0-65	This study
Lake Constance	450	93	Frenzel et al. (1990)
Lake Washington	250-350	50	Kuivila et al. (1988)
Vilhelmsborg Sø	97-6250	10-90	King (1990)

The aerobic methane oxidation rates measured in these marine sediments were similar to the rates reported from freshwater sediments. There were no real correlation between the methane concentration in the sediment and the measured rates. This missing relationship may be caused by the fact that the aerobic methane oxidation rates showed a large seasonal variation (Fig.4).

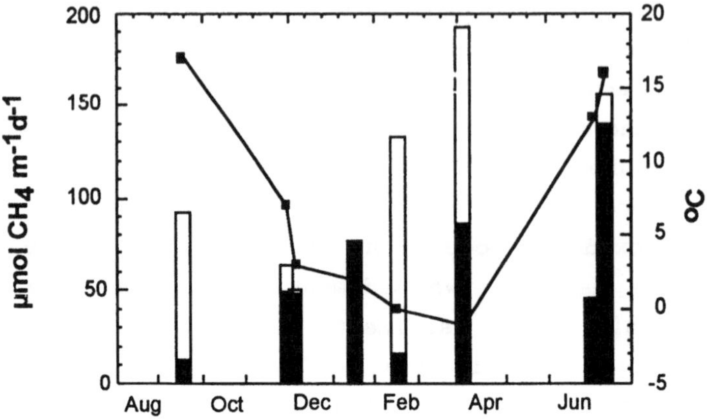

Fig.4 *In-situ* sediment water flux rates. The full length of the bar represents the potential methane flux under anaerobic conditions. The filled part of the bar represents net methane flux measured under oxic conditions. The difference between the bars represents the amount of methane oxidized.

The amount of methane oxidized varied from less than 1% to a nearly complete oxidation of the sediment-water flux. Based on a yearly average, approximately 50% of the sediment-water flux was oxidized and the rest escaped into the water column. The *in-situ* methane oxidation rate did not show any correlation with the *in-situ* temperature (Fig. 5A), and some of the highest rates were measured during the winter months. Nevertheless, surface sediment collected at an *in-situ* temperature of 11°C and incubated at different temperatures did show a temperature effect (Fig.5B). The rates measured could not be fitted with the Arhenius equation and the results show that the temperature had to be higher than 20 °C before any significant temperature stimulation was observed. This is a bit strange, since high *in-situ* rates could also be measured during the winter months (Fig.5). This is based on only one experiment, so clearly more experiments are needed before the temperature dependence of anaerobic methane oxidation in marine sediments can be determined.

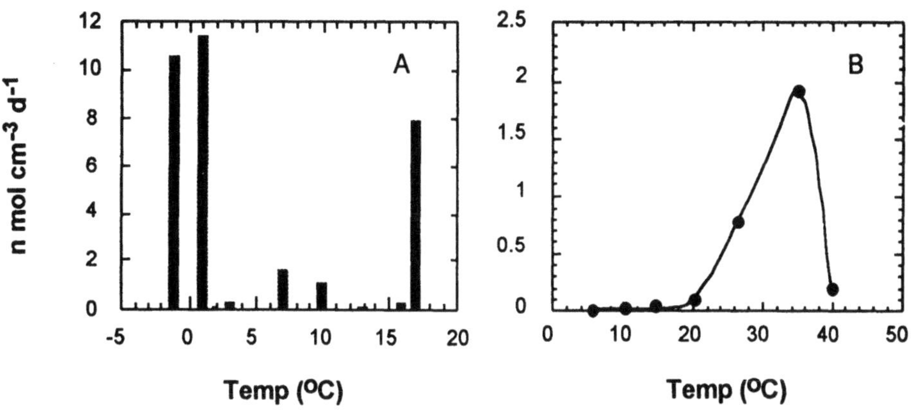

Fig.5. A. *In-situ* rates of aerobic methane oxidation vs. temperature. B. Temperature effect on surface sediments collected at an *in-situ* temperature of 11°C.

The large seasonal variation of aerobic methane oxidation could indicate that the methane oxidation in marine sediments is affected more by oxygen availability than by temperature. Oxygen has been found to be an important factor for regulation of aerobic methane oxidation in freshwater sediments (King, 1990; Frenzel.et al., 1990).

The potential for aerobic methane oxidation in the marine sediments studied was very high and a potential aerobic methane oxidation activity could be detected down to 6 cm in the sediment (Fig. 6). The highest potential rate was observed at the sediment surface and decreased sharply with depth down to 3 cm in the sediment. The potential methane oxidation activity showed no apparent lag phase indicating that the marine methane oxidizers are able to survive and keep the methane oxidation activity under anaerobic conditions.

Again, this resembles the findings from freshwater sediments, where anoxia tolerant methane oxidizing activity has also been observed (King, 1990). The normal oxygen penetration in these near-shore sediments is only a few mm, but due to generally lower microbial activity during the winter months (Sørensen et al.1979). Thus, more oxygen may be available for methane oxidation and the high oxidation rates measured in the winter could therefore be a result of increased oxygen availability.

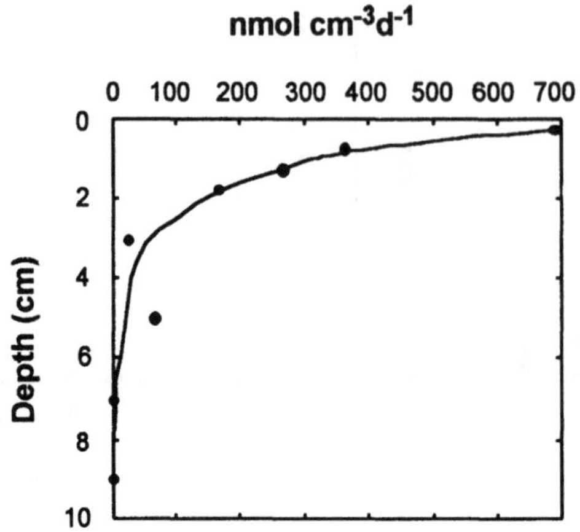

Fig.6. Potential aerobic methane oxidation rates measured in a near-shore marine sediment.

Experiments, where the oxygen transport into the sediment was changed from molecular diffusion to turbulent transport showed that the methane oxidation activity in the sediment increased from a 5% oxidation of the sediment-water flux to a complete oxidation of this flux (Güss, unpublished).

The importance of aerobic methane oxidation in marine sediments is probably significantly less than anaerobic methane oxidation. However, the few existing results indicate that aerobic methane oxidation might be responsible for the removal of as much methane as emitted to the atmosphere.

Methane oxidation in the water column

The average methane concentration in the open ocean surface waters has, based on 63 measurements, been estimated to be 2.5 ± 0.8 nmol l^{-1} (STP) (Lambert and Schmidt, 1993). The small variation of this number indicate the surface concentration of methane in the open oceans is quite uniform throughout the world. The reported methane concentrations from the surface water of the coastal zone show a much greater variation with concentrations ranging from 2.5 up to 1600 nmol l^{-1} (STP) (Lambert and Schmidt, 1993, Sansone and Martens; 1978; Fenchel et al. 1995)

Table 7. Water column methane oxidation rates.

	nmol l^{-1} d^{-1}		
	Aerob	Anaerob	Reference
Mariager Fjord	6	1-2	Fenchel et al. (1995)
Black Sea	0.001 - 0.1	1.6-19	Reeburgh et al. (1992)
Framvaren Fjord		0-11000	Lidstrom (1983)
Cape Lookout Bight	4 -17		Sansone and Martens (1977)
Bering Sea inshore	0.03 - 0.15		Griffiths et al. (1982)
Bering Sea offshore	0.01 - 0.07		Griffiths et al. (1982)
Cariaco Trench		0.5	Ward et al. (1987)
Deep Ocean	0.0004		Scranton and Brewer (1978)

Both aerobic and anaerobic methane oxidation has been observed in the water column (Table 7). Most of the published rates originate from fjords and sheltered areas with anoxic bottom water. These sites normally have methane concentrations in the μmolar range which are much higher than the nmolar methane concentration normally observed in oxic water columns. This may also explain why the reported anaerobic methane oxidation rates generally are much higher than the aerobic methane oxidation rates.

Only the reference from the Bering Sea deal with the more open coastal marine area, where the methane concentration in the water column varied from 4 - 43 nmol l^{-1}, which fall into the concentration range reported from coastal marine areas. If these rates are representative of the coastal zone, it can then be estimated that the turnover time for methane in the water column is in the order of years. Therefore, it is probably fair to say that methane oxidation in the water column is relative insignificant, but as the discussion is "data poor", more results are needed in order to estimate the role of methane oxidation in the water column of coastal marine areas.

Methane budget in coastal marine sediments.

Even though there is only a very limited number of data available on methane production and oxidation from the coastal marine environment, I will attempt to make a mass balance for methane in the coastal area (Fig. 7). The budget should only be taken as an indication of the methane dynamics in coastal marine areas and much more data are needed in order to make a final budget.

The source strength of methane is virtually unknown, but as oversaturation of methane is commonly observed below the sulfate reduction zone in marine sediments, methane production has to take place at the depths below the sulfate reduction zone. In order to sustain the saturating methane concentrations, the methane production at these depths has to be higher than or equal to the upward methane flux and thus, the minimum methane production can be calculated from the observed methane concentration profiles. Methane profiles have been reported from a variety of sediments around the world (Reeburgh, 1980; Devol, 1983; Iversen and Jørgensen, 1985; Alperin and Reeburgh, 1985).

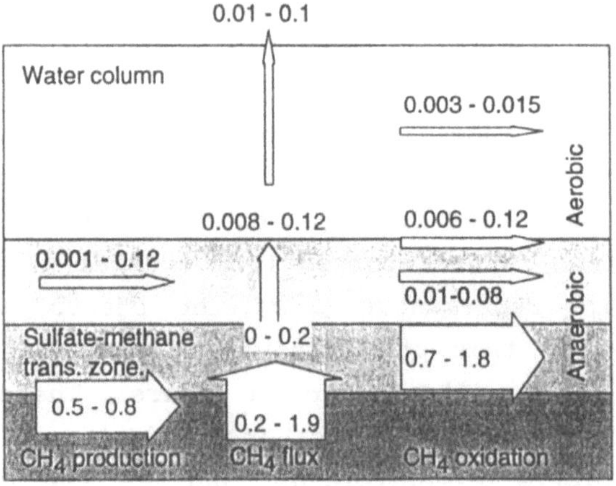

Fig. 7. Methane budget for a coastal marine sediment.All rates are in mmol $m^{-2}d^{-1}$; the numbers are explained in the text.

The published methane profiles are rather similar and the calculated upward methane flux ranges from 0.19-1.8 mmol $m^{-2}d^{-1}$, which can be used as a first estimate over methane production. The above mentioned references had also measured anaerobic methane oxidation at the sulfate methane transition zone and the integrated rate of methane oxidation at this zone is 0.7-1.9 mmol $m^{-2}d^{-1}$ (Table 4). The measured methane oxidation at the transition zone accounts for an oxidation of 80 to 200% of the upward methane flux. To balance the observed methane oxidation rates, a methane production of 0.5-0.8 mmol $m^{-2}d^{-1}$ has to take place at or immediately below the sulfate-methane transition zone.

The methane production rates at the sulfate containing depths are probably very low due to the substrate competition with the sulfate reducers. But, assuming steady state in the sediment, methane production has at least to be in the order of the integrated methane oxidation of 0.01-0.08 mmol $m^{-2}d^{-1}$ measured in the sulfate reduction zone (table 4).

An integrated methane production rate of 0.12 mmol $m^{-2}d^{-1}$ has been found in the sulfate reduction zone (0-120 cm) in a Danish coastal sediment (Kuivila, pers. comm.)

Aerobic methane oxidation in marine sediments has only been measured at very few stations in Danish near shore sediments. The rates show a high seasonal variability (Fig.4), so I have used the yearly methane oxidation rate to estimate a daily rate of 0.008-0.12 mmol $m^{-2}d^{-1}$. On a yearly basis, aerobic methane oxidation removed approximately 50% of the potential sediment-water flux, therefore I have estimated the methane release from the sediment to the water column to be in the same order as the aerobic methane oxidation rate.

Finally, some of the released methane is oxidized in the water column. As already mentioned, the sites with anoxic bottom waters probably represent extreme situations and in a more general budget it is reasonable to assume that there is no anaerobic methane oxidation in the water column. Aerobic methane oxidation has been measured in several water columns (Table 7), but the inshore rates measured in the Bering Sea are probably the ones that represent rates from the coastal zone best. In order to convert these rates to areal rates I have used an average water depth of 100 m yielding a methane oxidation in the water column of 0.003 to 0.015 mmol $m^{-2}d^{-1}$. Thus, methane oxidation in the water column removes only a small fraction of the methane released to the water column from the sediment and in this budget methane emission to the atmosphere is from 0.01-0.1 mmol $m^{-2}d^{-1}$ (Fig.8).

The coastal zone occupies app. 2.5 x 10^{13} m^2 , and using my estimated methane emission, the coastal zone would emit between 0.4 to 16.1 Tg CH_4 y^{-1}. This estimate is surprisingly close to Ehhalt's (1974) estimate for the coastal zone (Table 2), but much lower than estimates by Hovland et al. (1993) and Lamberth and Smith (1993). Both estimates included emission from natural methane seeps and organic rich sediments. I have not included the organic rich sediments in my budget, but emission rates as high as 60 mmol m^{-2} d^{-1} have been measured in areas with organic rich sediments (Martens and Klump, 1980). This rate is 500 times higher than the emission rate I have used, so even if the organic rich sediments only represent a very small fraction of the coastal marine area, they could be very significant with respect to methane emission.

The budget shown in Table 8 shows that anaerobic methane oxidation is the dominant methane sink in marine sediments. Nearly all the methane produced in the sediment is oxidized in the anaerobic zone and only a very small fraction of the methane produced reaches the aerobic surface layer (Table 8)

Table 8. Calculated methane production and methane oxidation in the coastal zone using an area of 2.5 x 10^{13} m^2 for the coastal zone.

	Tg y^{-1}	% oxidized
Methane production	120-248	
Anaerob methane ox.	104-245	87 - 99
Aerob. methane ox. sediment	1.1-17.5	1 - 9
Aerob. methane ox. water column	0.4-2.2	0.3-0.8

The importance of anaerobic methane oxidation becomes less significant (may only account for 1 to 25 % of the produced methane), if the sediment receives a high organic loading and from these areas a large increase in methane emission could be anticipated.

More data are needed on the relation between carbon loading, methane production and methane oxidation before more reliable estimates of the methane dynamics in the coastal marine area can be made.

References

Alperin MJ, Reeburgh WS (1985) Inhibition experiments on anaerobic methane oxidation. Appl Environ Microbiol 50:940-945

Chanton JP, Martens CS, Kelley CA (1989) Gas transport from methane-saturated, tidal freshwater and wetland sediments. Limnol Oceanog 34:807-819

Cicerone RJ, Oremland RS (1988) Biogeochemical aspects of atmospheric methane. Global Biogeochem Cycles 2:299-327

Crill PM, Martens CS (1983) Spatial and temporal fluctuations of methane production in anoxic coastal marine sediments. Limnol Oceanogr 28(6):1117-1130

Crill PM, Martikainen PJ, Nykanen H, Silvola J (1994) Temperature and N fertilization effects on methane oxidation in a drained peatland soil. Soil Biol Biochem 26: 1331-1339

Devol AH (1983) Methane oxidation rates in the anaerobic sediments of Saanich inlet. Limnol Oceanogr 28(4):738-742

Devol AH, Anderson JJ, Kuivila K, Murray JW (1984) A model for coupled sulfate reduction and methane oxidation in the sediments of Saanich inlet. Geochim Cosmochim Acta 48:993-1004

Ehhalt DH (1974) The atmospheric cycle of methane. Tellus 26:58-70

Fenchel, Bernard TC, Esteban G, Finlay BJ, Hansen PJ, Iversen N (1995) Microbial diversity and activity in a Danish Fjord with anoxic deep water. Ophelia (In press)

Frenzel P, Thebrath B, Conrad R (1990) Oxidation of methane in the oxic surface layer of a deep lake sediment (Lake Constance). Fems Microbiol Ecol 73(2):149-158

Griffiths RP, Caldwell BA, Cline JD, Broich WA, Morita RY (1982) Field observation of methane concentrations and oxidation rates in the southeastern Bering Sea. Appl Environ Microbiol 44:435-446

Hansen LB, Finster K, Fossing H, Iversen N (1995) Effect of sulfate and molybdate addition on anaerobic methane oxidation in sulfate depleted marine sediments. (in prep)

Hoehler T, Alperin MJ, Albert DB, Martens CS (1994) Field and laboratory studies of methane oxidation in an anoxic marine sediment: Evidence for a methanogen-sulfate reducer consortium. Global Biogeochem Cycles 8: 451-463

Hovland M, Judd AG, Burke RA (1993) The global flux of methane from shallow submarine sediments. Chemosphere 26(1-4):559-578

Iversen N, Blackburn TH (1981) Seasonal rates of methane oxidation in anoxic marine sediments. Appl and Environ Microbiol 41(6):1295-1300
Iversen N, Jørgensen BB (1985) Anaerobic methane oxidation rates at the sulfate-methane transition in marine sediments from Kattegat and Skagerrak (Denmark). Limnol Oceanogr 30:944-955
Iversen N, Jørgensen BB (1993) Diffusion coefficients of sulfate and methane in marine sediments: Influence of porosity. Goechim Cosmochim Acta 57:571-578
Iversen N, Oremland RS, Klug MJ (1987) Big Soda Lake (Nevada). 3. Pelagic methanogenesis and anaerobic methane oxidation. Limnol Oceanogr 32(4):804-814
King GM (1990) Dynamics and control of methane oxidation in a Danish wetland sediment. FEMS Microbiol Ecol 74: 309-324
Kuivila KM, Murray JW, Devol AH, Lidstrom ME, Reimers CE (1988). Methane cycling in the sediments of Lake Washington. Limnol and Oceanogr 33:571-581
Lambert G, Schmidt S (1993) Reevaluation of the oceanic flux of methane - uncertainties and long term variations. Chemosphere 26(1-4):579-589
Lidstrom ME (1983) Methane consumption in Framvaren, an anoxic marine fjord. Limnol Oceanogr 28:1247-1251
Martens CS, Berner RA (1977) Interstitial water chemistry of anoxic Long Island Sound sediments. 1. Dissolved gases. Limnol Oceanogr 22: 10-25
Martens CS, Klump JV (1980) Biogeochemical cycling in a organic-rich coastal marine basin 1. Methane sediment-water exchange processes. Geochim Cosmochim Acta 44:471-490
Nesbit SP, Breitenbeek GA (1992) A laboratory study of factors influencing methane uptake by soils. Agric Ecosyst Environ 41: 39-54
Reeburgh WS (1980) Anaerobic methane oxidation: Rate depth distributions in Skan Bay sediments. Earth and Planet Sci Lett 47: 345-352
Reeburgh WS, Ward BB, Whalen SC, Sandbeck KA, Kilpatrick KA, Kerkhof LJ (1992) Black Sea methane geochemistry. Deep-Sea Res (Black Sea Issue), 38:1189-1210
Sansone FJ, Martens CS (1978) Methane oxidation in Cape Lookout Bight, North Carolina. Limnol Oceanogr 23(2):349-355
Scranton MI, Brewer PG (1978) Consumption of dissolved methane in the deep ocean. Limnol Oceanogr 23:1207-1213
Sørensen J, Jørgensen BB, Revsbech NP (1979) A comparison of oxygen, nitrate, and sulfate respiration in coastal marine sediments. Microbial Ecology 5:105-115
Ward BB, Kilpatrick KA, Novelli PC, Scranton MI (1987) Methane oxidation and methane fluxes in the ocean surface layer and deep anoxic waters. Nature 327:226-229
Zehnder AJB, Brock TD (1980) Anaerobic methane oxidation: Occurrence and ecology. Appl Environ Microbiol 39:194-204

Microbial Controls of Methane Oxidation in Temperate Forest and Agricultural Soils

[1]PA Steudler, RD Jones[2], MS Castro[3], JM Melillo[1], and DL Lewis[4]

[1]The Ecosystems Center, Marine Biological Laboratory, Woods Hole, MA. U.S.A. [2]Drinking Water Research Center, Florida International University, Miami, FL [3]Appalachian Environmental Laboratory, University of Maryland, Frostburg, MD [4]United States Environmental Protection Agency, Environmental Research Laboratory, Athens, GA

Introduction

The tropospheric accumulation of the radiatively active gas methane (CH_4) has been well-documented with annual increases of approximately 1% measured over the past several decades (Prather et al. 1995, Watson et al. 1990, 1992). Understanding the interactions between the biosphere and atmosphere is necessary to understand how future human activities will affect the global atmospheric CH_4 budget.

Methane emissions from both biogenic and anthropogenic sources have been documented (Watson et al. 1992). Some studies also report how human activities affected the emissions (Watson et al. 1990 and references cited therein). The annual rate of CH_4 oxidation by OH radicals in the troposphere is well-known but the estimated rates of CH_4 removal by aerobic soils (Prather et al. 1995) vary by a factor of 3 (15-45 Tg CH_4). Methane consumption by temperate forest soils and the factors controlling the fluxes have been well documented (Steudler et al. 1989, Born et al. 1990, Yavitt et al. 1990, Crill 1991, King and Adamsen 1992, Nesbit and Breitenbeck 1992, Dorr et al. 1993, Koschorreck and Conrad 1993, Castro et al 1994b, 1995). However, our understanding of the effects of nitrogen fertilization, changes in the rates of nitrogen turnover, and land-use (Steudler et al. 1989, 1991, Keller et al. 1983, Mosier et al 1991, Ojima et al.

NATO ASI Series, Vol. I 39
Microbiology of Atmospheric Trace Gases
Edited by J. Colin Murrell and Donovan P. Kelly

1993) on the microbial communities responsible for the oxidation of CH_4 is still evolving (Bender and Conrad 1993, 1994, King 1993, King and Schnell 1994ab, Schnell and King 1994, Castro et al. 1994a, 1995). One of the key questions in understanding the mechanism of soil CH_4 consumption is the identification of the soil microorganisms responsible for the oxidation.

In this paper, we describe our adaptation of a rapid and sensitive laboratory assay that can be used to distinguish between the two major microbial communities, methanotrophs and nitrifiers, believed to be responsible for the observed consumption of CH_4 by soils. In addition, we describe how this technique can be used to explain the large spatial variation in CH_4 consumption by temperate forest soils.

Site Descriptions

All study sites are located in the southeastern or west-central regions of Massachusetts, USA. Site characteristics are summarized in Table 1.

Two of the forest study sites are in Prospect Hill Tract at the Harvard Forest in Petersham. This tract has a ~65-year-old red pine (*Pinus resinosa* Ait.) plantation and an adjacent ~80-year-old mixed hardwood stand dominated by black oak (*Quercus velutina* Lam.). Both stands are on mor soils. These stands have been the focus of a long-term chronic nitrogen fertilization experiment since 1988. In 3-30x30 m areas within each of these stands, we established 36-5x5 m control, low (50 kg N/ha*yr) and high (150 kg N/ha*yr) NH_4NO_3 fertilized subplots. We have been measuring a number of parameters in these plots including soil trace gas fluxes and nitrogen dynamics (Steudler et al. 1989, Bowden et al. 1990, 1991, Aber et al. 1993, Castro et al. 1994a, 1995).

The sugar maple (*Acer saccharum* Marsh) site is in the Mount Toby Forest near the towns of Sunderland and Leverett in west-central Massachusetts owned by the University of Massachusetts. The forest is an even-aged mature stand that had been in farmland or pasture from the mid-1800s until the early 1900s (Boone 1992). This is a mull soil about 40-50cm deep most likely of aeolian origin.

Table 1. Site characteristics of the five locations used in this study.

Location	Vegetation	Soil type/Series	Texture	Bulk density (g/cm^3)	pH
Harvard Forest/ Prospect Hill Tract	Red pine plantation	Spodosol/Gloucester Exotic Haplothods	Sandy loam	0.90[1]	3.2[3]
Harvard Forest/ Prospect Hill Tract	Mixed hardwood	Spodosol/Gloucester Exotic Haplothods	Sandy loam	0.94[1]	3.3[3]
Mount Toby Forest	Sugar maple	Mull/Merrimac Typic Dystrochrept	Sandy loam	0.91[2]	5.4[2]
West Falmouth	Black locust	Spodosol	Sandy	0.82[1]	3.8[3]
South Deepfield	Cornfield	Alluvium Typic Udifluvent	Hadley silt loam	1.12[2]	6.8[2]

1 Upper 10cm mineral soil
2 0-15cm mineral soil
3 Forest floor

The black locust stand is on a tract of private land in West Falmouth on Cape Cod. The stand is even aged (>45 years) and dominated (78% of the basal area) by *Robinia pseudoacacia* with black cherry (*Prunus serontina* - 22%). The presence of an Ap layer indicated that this site had previously been in agriculture (Melillo unpublished).

The cornfield (*Zea mays* L.) site is on the Connecticut River floodplain in South Deerfield and has been managed for corn silage by the University of Massachusetts for the last 25 years (Boone 1990). Methane fluxes and soil samples were taken within the corn row. The cornfield was fertilized with 103 kg N/ha (NPK) in early May and again with 52 kg N/ha as urea in late June. Measurements of soil NH_4^+ and NO_3^- pools over the growing season suggested that nearly all the applied N is subsequently nitrified (Boone 1990).

Field Measurements and Methods

In the first experiment, three soil cores (0-5cm depth) of the organic horizon and upper mineral soil were collected in July and/or August 1989 at all sites. Cores were composited, roots and rocks removed by hand and the soils homogenized. Soil subsamples were taken for the CH_4 oxidizer assay and gravimetric soil moisture determinations. In the second experiment, three soil cores to a depth of 15cm were collected in August 1992 at the Harvard Forest pine and hardwood control, high-N fertilized plots and locust stand. Cores were separated into the organic layer and the upper 10cm mineral soil. The separated layers were composited, roots and rocks removed and the soils homogenized. Soil subsamples of the separate composited layers were taken for CH_4 oxidizer assay and soil moisture determinations.

Methane fluxes were measured on the same date as the soil samplings. Four static chambers were used at all sites except the Harvard Forest plots where three chambers were used (Steudler et al 1989, Bowden et al 1990). At the sugar maple and cornfield sites, fluxes were measured five times over the day at 0700, 1100, 1500, 1900 and 2300 hours. At the locust stand and Harvard Forest plots, methane fluxes were measured six times over the day at 0500, 0900, 1300, 1700, 2100 and 0100 hours in 1989 and four times at 0600, 1000, 1400 and 1800 hours in 1992. Soil temperatures at 0-2.5 and 2.5-5 cm were measured during each incubation.

Net nitrogen mineralization and net nitrification rates were measured at all sites using the in situ buried bag technique (Nadelhoffer et al. 1983, Pastor et al. 1984). We averaged the Harvard Forest net N-mineralization and nitrification rates for the years 1988 and 1990 (Aber et al. 1993) and 1991 and 1993 (Magill et al. submitted) to calculate the net N mineralization and nitrification rates for the years 1989 and 1992, respectively.

Assays of CH_4 oxidizer activity were determined by using methods similar to those described by Jones and Morita (1983a,b) for measuring methane and carbon monoxide oxidation rates. Soils were stored at field moisture in the dark at <25 ^{o}C until use. Soils were stored generally for less than four days after collection before incubations were begun. For all soils 1.0 g wet weight (at field soil moistures) was weighed into a 60 ml serum bottle. A subset of bottles containing the soil received 25 ml sterile distilled water which represented saturated conditions. Bottles were then purged for 3 min with room air, at a flow rate of 100 ml/min. to remove any CH_4 and CO in the soil or soil water mixture. Ambient concentrations of these gases in the atmosphere is extremely low compared to the assay conditions, it was assumed that there would be a negligible dilution effect due to non-radiolabeled CH_4 and CO present in the air. Bottles were then sealed with serum stoppers and either 0.5 ml of $^{14}CH_4$ diluted in nitrogen (1.0 μCi/ml; specific activity, 59 mCi/mmole; Amersham Corp.) or 0.5ml of ^{14}CO diluted in nitrogen (0.5 μCi/ml; specific activity 56 mCi/mmole: Amersham Corp.) was injected into the headspace. All bottles were prepared in triplicate and acid controls were run with each experiment. Bottles were then incubated for 2 hours for CO, or 24 hours for CH_4 on a rotary shaker at 100rpm. All samples were incubated at 25 ^{o}C in the dark. Bottles used to determine the N-serve (NS) differential (for CO oxidation; see Jones et al., 1984) received either 25 μl of 100 μg/μl NS (Chem Services, Westchester, PA) in DMSO or 25 μl of DMSO as controls. These bottles were stoppered as before but were pre-incubated for 1 hour on a rotary shaker at 100 rpm prior to the addition of ^{14}CO and incubated as described above. After incubation, the reaction was terminated by the addition of 1.0 ml of 5.0 N NaOH with a syringe through the stopper. The serum bottles were then shaken for 1 hour at room temperature to permit the labeled CO_2 to be absorbed into solution. The stoppers were then removed and the bottles shaken for an additional 30 min in an exhaust hood to remove the remaining labeled CH_4 or CO. Labeled CO_2 was released, trapped and assayed as previously described by Griffiths et al (1982). Methane and CO oxidation rates were calculated using solution equilibrium concentrations based on headspace mixing ratios and the appropriate Bunsen solubility coefficient (Jones et al. 1984). Methane and CO oxidation rates were expressed on a per g dry weight basis at a concentration of 11.93 nM for CH_4 and 2.23 nM for CO. These values were chosen since they allowed comparison between this work and previous studies. The ratio of CH_4 oxidation to CO oxidation was calculated using these values. Only the NS inhibited CO oxidation was considered for these calculations. Based on the results from pure culture incubations of methanotrophs and ammonium oxidizers, $^{14}CH_4/^{14}CO$ ratios of >0.05 suggest that CH_4 oxidizers were the dominant group while ratios of <0.05 suggest that NH_4 oxidizers were the primary group (Jones et al. 1984). If ratios were near the 0.05 value, we assigned the oxidation of CH_4 to both communities.

We tested the assumption that $^{14}CH_4$ oxidation rates were constant during the 24-hour incubation period for our soil samples. We chose soils from the Harvard Forest hardwood control plot, a site with very high CH_4 consumption rates, and the pine high-N plot, a site with a much lower CH_4 consumption rates. Soils from the organic and mineral horizons were incubated as described above and the rate of $^{14}CH_4$ oxidation measured over a time course of 6, 12, 24 and 48 hours. We found that the rate of CH_4 oxidation was constant over time for both soils for the 48-hour incubation period (data not shown).

Results

Soil temperatures at 0-2.5cm depth ranged from approximately 14 to 18 ^{o}C for all sites during July and August 1989 samplings. Soil moistures also varied little between sites during this time. Chamber CH_4 consumption rates for the 1989 experiment varied more than 4-fold (-0.118 to -0.028 mg CH_4-C/m^2*h) across the forest sites (Table 2). Highest consumption rates were measured in the Harvard Forest control plots and the lowest in the Locust stand. We measured 15% and

Table 2. Results from the 1989 field measurements and laboratory oxidation experiments.

Site	Soil Moisture (%)	Chamber CH_4 consumption (mg CH_4-C/m^2 h)	$^{14}CH_4$ oxidation (nmols/gdw/h)	^{14}CO oxidation (nmols/gdw/h)	Ratio $^{14}CH_4/^{14}CO$ oxidation	Dominant Bacteria
Harvard Forest						
Hardwood						
Control	34.0	-0.115 (.011[1])	0.993	2.449	0.405	CH_4 oxidizers
High N	40.0	-0.097 (.014)	0.757	3.071	0.247	CH_4 oxidizers
Pine Plantation						
Control	28.0	-0.118 (.010)	1.027	2.303	0.446	CH_4 oxidizers
High N	22.0	-0.068 (.013)	0.108	2.269	0.048	NH_4 & CH_4 oxidizers
Sugar Maple (July)	*	*	0.079	2.397	0.033	NH_4 oxidizers
Sugar Maple (Aug)	23.7	-0.067 (.009)	0.087	1.689	0.052	NH_4 & CH_4 oxidizers
Locust	31.2	-0.028 (.003)	0.001	1.560	0.0008	NH_4 oxidizers
Cornfield (July)	*	*	0.0004	1.377	0.0003	NH_4 oxidizers
Cornfield (August)	27.7	+0.0002 (.0001)	0.002	1.190	0.0015	NH_4 oxidizers

1 ± standard error of mean

* not measured

41% lower rates of CH_4 consumption between the hardwood and pine fertilized plots and the corresponding control plots, respectively. There were even larger reductions in the consumption rates in the sugar maple (42%) and locust (76%) stands compared to the hardwood control. The cornfield was a slight source of CH_4 (+0.0002 mg CH_4-C/m^2*h) to the atmosphere.

Corresponding laboratory $^{14}CH_4$ oxidation rates paralleled those measured using chambers. The $^{14}CH_4$/^{14}CO ratios suggested that CH_4 oxidizers were the dominant bacteria at the sites with the highest oxidation rates and NH_4 oxidizers dominated at the sites with the lowest oxidation rates. The pine high-N plot and the sugar maple stand had intermediate $^{14}CH_4$/^{14}CO ratios suggesting mixed populations were responsible for the oxidation. We made two sets of laboratory assays in July and August in the sugar maple stand and cornfield. Comparison of the $^{14}CH_4$, ^{14}CO oxidation rates and $^{14}CH_4$/^{14}CO ratios from these two dates indicated that the assay gave reproducible results.

The results from the second set of measurements made in August 1992 at the Harvard Forest hardwood and pine control, high-N plots and the locust stand are presented in Table 3. Methane consumption rates showed the same pattern as observed in 1989 with the highest consumption rates measured in the Harvard Forest control plots. There was an even larger decrease (61%) in the CH_4 consumption rate in the pine high-N plot compared to the control. We also investigated which microbial communities were responsible for CH_4 oxidation by soil horizon. Ammonium oxidizers were generally the dominant bacteria in the organic horizon but had $^{14}CH_4$ oxidation rates that were 4 to 15-times lower than the rates measured in the mineral horizon (Table 3). Methane oxidizers were the dominant bacteria in the mineral soil at all sites but CH_4 oxidation rates were more than 6-times lower in the more highly nitrifying pine high-N and locust stand.

Annual net nitrogen mineralization rates in the Harvard Forest control plots were between 70 and 80 kg N/ha during 1989 and 1992 (Table 4). The pine high-N plot and sugar maple stand had the highest mineralization rates up to 170 kg N/ha. Net nitrification rates were always very low, <1 kg N/ha, in the hardwood control plot while in the pine control plot rates were much higher, about 19 kg N/ha. The highest nitrification

Table 3. Results from the 1992 field measurements and laboratory oxidation experiments.

Site		Soil Moisture %	Chamber CH_4 consumption (mg CH_4-C/m^2 h)	$^{14}CH_4$ oxidation (nmols/gdw/h)	^{14}CO oxidation (nmols/gdw/h)	Ratio $^{14}CH_4/^{14}CO$ oxidation	Dominant Bacteria
Harvard Forest							
Hardwood							
Control	O[1]	57.6	-0.171 (.011)	0.018	0.698	0.0258	NH_4 oxidizers
	M[2]	27.4		0.131	0.320	0.4094	CH_4 oxidizers
High N	O	134.6	-0.153 (.024)	0.022	0.795	0.0277	NH_4 oxidizers
	M	36.5		0.149	0.361	0.4127	CH_4 oxidizers
Pine Plantation							
Control	O	96.1	-0.195 (.025)	0.060	0.667	0.0899	CH_4 oxidizers
	M	22.2		0.041	0.286	0.1433	CH_4 oxidizers
High N	O	94.4	-0.076 (.021)	0.004	0.782	0.0051	NH_4 oxidizers
	M	22.1		0.016	0.173	0.0925	CH_4 oxidizers
Locust	O	65.1	-0.025 (.015)	0.001	2.178	0.00046	NH_4 oxidizers
	M	22.2		0.015	0.232	0.0646	CH_4 oxidizers

1 organic horizon
2 upper 10cm mineral soil

Table 4. Annual net N mineralization, nitrification, relative % nitrification and CH_4 consumption rates.

Site	Net N-mineralization (KgN/ha·yr)	Net nitrification (KgN/ha·yr)	Relative % Nitrification	Annual CH_4 consumption[3] ($mg\ CH_4\text{-}C/m^2 \cdot h$)
Harvard Forest				
Hardwood (1989)[1]				
Control	70.44	0.02	< .1	-0.110
High-N	96.02	3.37	3.50	-0.070
Hardwood (1992)[1]				
Control	76.70	1.03	1.34	-0.130
High-N	170.29	3.84	2.25	-0.080
Pine (1989)[1]				
Control	80.21	18.26	22.80	-0.100
High-N	129.12	32.99	25.55	-0.060
Pine (1992)[1]				
Control	77.90	19.53	25.07	-0.110
High-N	99.29	51.39	51.76	-0.040
Sugar maple (1987)[1]	107.9	49.8	46.1	—
Locust (1981)[1]	40.0	40.0	100	-0.025[2]
Cornfield (1987)[1]	—	—	> 90	—

1 Annual measurments made during that year
2 Steudler unpublished (1991)
3 Castro et al. 1995

rates, up to 51 kg N/ha, were measured in the sugar maple and in the pine high-N plot during 1992. We calculated a relative % nitrified (nitrification rate/mineralization rate) for all sites and found that in the locust stand and cornfield nearly 100% of the amount of N-mineralized was nitrified. The pine high-N site had a value of 25% in 1989 which increased to 51% in 1992. The sugar maple site had an intermediate value of about 46%.

Discussion

All the temperate forest sites we examined in this study consumed atmospheric CH_4, but at rates that varied by a factor of 7.8 (-0.195 to -0.025 mg CH_4/m^2*h). This large range in consumption rates is not uncommon. For example, Koschorreck and Conrad (1993) and Castro et al (1995) reported a 10-fold variation in CH_4 consumption for temperate forest soils in Germany and the United States. Koschorreck and Conrad suggested that the 10-fold range in spatial variation could be explained by soil texture alone. In contrast, Castro et al (1995) suggested that the activity of the microbial communities that oxidize atmospheric CH_4 is primarily responsible for the spatial variations in CH_4 consumption. To better explain the spatial variations in CH_4 consumption by temperate zone soils, we used a sensitive laboratory assay to identify the relative activities of the microbial communities that oxidize CH_4 in soils that were located along a natural gradient in soil nitrogen turnover and in soils that were fertilized with nitrogen.

The Harvard Forest control plots, sugar maple and black locust stands form a natural gradient in nitrogen turnover that ranges from <1 to 100% relative % nitrification (Table 4). The Harvard Forest control plots had the lowest turnover rates (<1 to 25%), the sugar maple stand had intermediate rate (ca. 50%) and the black locust stand had the highest rate (100%). We found that across this gradient, from low to high N turnover, there was a 5-fold decrease in the annual CH_4 consumption rates (Table 4) and a shift in the relative activity of the soil microbial communities responsible for CH_4

consumption from those dominated by methanotrophs to those dominated by nitrifiers (Table 2). This pattern is consistent with the differences in the CH_4 oxidation rates measured in pure culture studies of these bacteria. Jones et al. (1984) reported that pure cultures of methanotrophic bacteria had 2 to 4 orders of magnitude greater CH_4 oxidation rates than pure cultures of nitrifying bacteria. Thus, these results support the idea that the rate of CH_4 consumption by soil is related to soil nitrogen turnover as hypothesized by Moiser et al. (1991), Mosier and Schimel (1991), Steudler et al. (1991) and Hutsch et al. (1993).

Results from our study also suggest that nitrogen fertilization can sometimes alter soil microbial activity and CH_4 consumption rates. We measured a reduction in CH_4 consumption and a shift in the relative activities of CH_4 oxidizing bacteria in the Harvard Forest pine high-N fertilized plot (Table 2 and 3). These observations are consistent with a study by Castro et al. (1994b) in fertilized and unfertilized slash pine plantations in Florida. They found a similar decrease in CH_4 consumption in the fertilized plots with a concomitant shift in the dominant group of CH_4 oxidizing bacteria from CH_4 oxidizers to ammonium oxidizers in the fertilized soils. Jones et al. (1984) reported that ammonium oxidizers were the dominant bacteria oxidizing CH_4 in soils of a fertilized garden in Oregon. We also found similar results in the fertilized cornfield where nitrifiers were the dominant bacteria.

In contrast, however, we did not obtain the same results from measurements made in the Harvard Forest hardwood stand. In this stand, we found no shift in the microbial communities that oxidize CH_4. The reason for this discrepancy may be related to fact that there are two phenomena affecting the magnitude of the suppression of CH_4 consumption in fertilized soils. One is the direct interaction of NH_4^+ from the fertilizer on the activity of the microbial communities. The other is the indirect effect of changing the rate of nitrogen turnover, specifically the relative % nitrified, over time through fertilization. We measured a nearly constant 38% reduction in CH_4 consumption in the hardwood high-N plot from 1989 through 1993 (Castro et al. 1995) and the relative %

nitrification was low and similar in both the control and fertilized plots during those years (Table 4). These results suggest that in this stand only the direct effect is operating because fertilization has not substantially altered the nitrogen turnover rate. However in the pine stand, we measured both a larger reduction in CH_4 consumption rates between 1989 and 1992 (40 to 64%) and also a doubling (25 to 51%) in the relative % nitrified (Table 4). Thus, in the pine high-N plot both the direct and indirect effects appear to be operating.

In summary, results from this study suggest that spatial variations in CH_4 consumption by temperate forest soils are related to site nitrogen turnover which in turn affects the activities of soil microbial communities. Soils with little or no nitrification have high CH_4 consumption rates because CH_4 oxidation is dominated by methanotrophs. In contrast, soil with high nitrification rates have low CH_4 consumption because CH_4 consumption is dominated by nitrifiers. This pattern exists along natural gradients in site fertility and can sometimes be induced by human fertilization and land-use activities.

Acknowledgments

We thank Michelle Miliefsky, Jon Chapman and Chris Kerfoot for their field and laboratory assistance. This research was supported by funds from the Ecosystems Studies (BRS-8919647)and Long Term Ecological Research programs of the National Science Foundation and by the United States Environmental Protection Agency (Cooperative agreement CR817734-010, Athens Environmental Laboratory).

References

Aber JD, Magill A, Boone R, Melillo JM, Steudler P, Bowden R, (1993) Plant and soil responses to chronic nitrogen additions at the Harvard Forest Massachusetts. Ecol Appl 3:156-166

Adamsen APS, King GM (1993) Methane consumption in temperate and subarctic forest soils: Rates, vertical zonation and response to water and nitrogen. Appl Environ Microbiol 59:485-490

Bender M, Conrad R (1993) Kinetics of methane oxidation in oxic soils. Chemosphere 26:687-696

Bender M, Conrad R (1994) Microbial oxidation of methane, ammonium and carbon monoxide, and turnover of nitrous oxide and nitric oxide in soils. Biogeochemistry 27:97-112

Boone RD (1990) Soil organic matter as a potential net nitrogen sink in a fertilized cornfield, South Deerfield, Massachusetts, USA. Plant and Soil 128:191-198

Boone RD (1992) Influence of sampling date and substrate on nitrogen mineralization: comparison of laboratory-incubation and buried-bag methods for two Massachusetts forest soils. Can J For Res 22:1895-1900

Born M, Dörr H, Levin I (1990) Methane consumption in aerated soils of the temperate zone. Tellus 42:(B) 2-8

Bowden RD, Steudler PA, Melillo JM, Aber JD (1990) Annual nitrous oxide fluxes from temperate forest soils in the northeastern United States. J Geophys Res 95:13,997-14,005

Bowden RD, Melillo JM, Steudler PA, Aber JD (1991) Effects of nitrogen additions on the annual nitrous oxide fluxes from temperate forest soil in the northeastern United States. J Geophys Res 96:9321-9328

Castro MS, Peterjohn WT, Melillo JM, Steudler PA, Gholz HL, Lewis D, (1994a) Effects of urea fertilization on the exchange of CO_2 N_2O and CH_4 between the atmosphere and soils in a Florida slash pine plantation. Can J For Res 24:9-13

Castro MS, Melillo JM, Steudler PA, Chapman JW (1994b) Soil moisture as a predictor of methane uptake by forest soils. Can J For Res 24: in press

Castro MS, Steudler PA, Melillo JM, Aber JD, Bowden RD (1995) Factors controlling atmospheric methane consumption by temperate forest soils. Global Biogeochem Cycles 9:1-10

Crill P (1991) Seasonal patterns of methane uptake and carbon dioxide relase by a temperate woodland soil. Global Biogeochem Cycles 5:319-334

Dörr H, Katruff L, Levin I (1993) Soil texture parameterization of the methane uptake in aerated soils. Chemosphere, *26*:(1-4), 697-713

Griffiths RP, Caldwell BA, Cline JD, Broich WA, Morita RY (1982) Field observations of methane concentrations and oxidation rates in the southeastern Bering Sea. Appl Environ Microbiol 44:435-446

Hutsch BW, Webster CP, Powlson DS (1993) Long-term effects of nitrogen fertilization on methane oxidation in soil of the broadwalk wheat experiment. Soil Biol Biochem 25:1307-1315

Jones RD, Morita RY (1983a) Carbon monoxide oxidation by chemolithotrophic ammonium oxidizers. Can J Microbiol 29:1545-1551

Jones RD, Morita RY (1983b) Methane oxidation by *Nitrosococcus oceanus* and *Nitrosomonas europaea*. Appl Environ Microbiol, 45:401-410

Jones RD, Morita RY, Griffiths RP (1984) Method for estimating in situ chemolithotrophic ammonium oxidation using carbon monoxide oxidation. Mar Ecol Prog Ser 17:259-269

Keller M, Goreau TJ, Wofsy SC, Kaplan WA, McElroy MB (1983) Production of nitrous oxide and consumption of methane by forest soils. Geophys Res Lett 10:1156-1159

King GM, Adamsen APS (1992) Effects of temperature on methane consumption in a forest soil and pure cultures of the *methanotroph methylomonas rebra*. Appl Environ Microbiol 58:2758-2763

King GM (1993) Ecological aspects of methane oxidation, a key determinant of global methane dynamics. In: *Advances in Microbial Ecology* (K.C. Marshall, ed.), Plenum Press, New York, pp. 431-468

King GM, Schnell S (1994a) Effect of increasing atmospheric methane concentration on ammonium inhibition of soil methane consumption. Nature 370:282-284

King GM, Schnell S (1994b) Ammonium and nitrite inhibition of methane oxidation by *Methylobacter albus* BG8 and *Methylosinus trichosporium* OB3b at low methane concentrations. Appl Environ Microbiol 60:3508-3513

Koschorreck M, Conrad R (1993) Oxidation of atmospheric methane in soil: Measurements in the field, in soil cores and in soil samples. Global Biogeochem Cycles 7:109-121

Magill AH, Aber JD, Hendricks JJ, Bowden RD, Melillo JM, Steudler PA Biogeochemical responses of forest ecosystems to simulated chronic nitrogen deposition. Submitted to Ecological Applications

Mosier AR, Schimel DS (1991) Influence of agricultural nitrogen on atmospheric methane and nitrous oxide. Chemistry and Industry 23: 874-877

Mosier A, Schimel D, Valentine D, Bronson K, Parton W (1991) Methane and nitrous oxide fluxes in native, fertilized, and cultivated grassland. Nature 350:330-332

Nadelhoffer KJ, Aber JD, Melillo JM (1983) Leaf-litter production and soil organic matter dynamics along a nitrogen mineralization gradient in Southern Wisconsin. Can J For Res 13:12-21

Nesbit SP, Breitenbeck GA (1992) A laboratory study of factors influencing methane uptake by soils. Agric Ecosyst Environ 41:39-54

Ojima DC, Valentine DW, Mosier AR, Parton WJ, Schimel DC (1993) Effect of land use change on methane oxidation in temperate forest and grassland soils. Chemosphere 26(1-4), 675-685

Pastor J, Aber JD, McClaugherty CA, Melillo JM (1984) Aboveground production and N and P cycling along a nitrogen mineralization gradient on Blackhawk Island, Wisconsin. Ecology 65:256-268

Prather M, Derwent R, Ehhalt D, Fraser P, Sanhueza E, Zhou X (1995) Other trace gases and atmospheric chemistry. In: *Climate Change 1994 Radiative Forcing of Climate Change and An Evaluation of the IPCC IS92 Emission Scenarios.*

(Houghton JT, Meira Filho LG, Bruce J, Hoesung Lee, Callander BA, Haites E, Harris N, Maskell K, eds.), Cambridge University Press, pp.74-126

Steudler PA, Bowden RD, Melillo JM, Aber JD (1989) Influence of nitrogen fertilization on methane uptake in temperate forest soils. Nature 341:314-316

Steudler PA, Melillo JM, Bowden RD, Castro MS (1991) The effects of natural and human disturbances on soil nitrogen dynamics and trace gas fluxes in a Puerto Rican wet forest. Biotropica 23(4a):356-363

Schnell S, King GM (1994) Mechanistic analysis of ammonium inhibition of atmospheric methane consumption in forest soils. Appl Environ Microbiol 60:3514-3521

Watson RT, Rodhe H, Oeschger H, Siegenthaler U (1990) Greenhouse gases and aerosols. In: *Climate Change The IPCC Scientific Assessment*, (Houghton JT, Jenkins GJ, Ephraums JJ, eds.), Cambridge University Press, New York, pp 1-40

Watson RT, Meira F, Sanhuez E, Janetos A (1992) Greenhouse gases: Sources and sinks and aerosols. In: *Climate Change 1992 The Supplementary Report to The IPCC Scientific Assessment*, (Houghton JT, Callander BA, Varney SK, eds.), Cambridge University Press, New York, pp 25-46

Yavitt JB, Downey DM, Lang GE, Sextone AJ (1990) Methane consumption in two temperate forest soils. Biogeochemistry 9:39-52

Microbial Degradation of Atmospheric Halocarbons

Ronald S Oremland
US Geological Survey
345 Middlefield Road
Menlo Park CA 94025 USA

Introduction

Halocarbons are present in the atmosphere at parts-per-trillion (ppt) mixing ratios and are represented by such substances as chlorofluorocarbons (CFCs), hydrochlorofluorocarbons (HCFCs), hydrofluorocarbons (HFCs), and methyl halides like methyl bromide (MeBr) and their further substituted halomethane analogues (e.g., dibromomethane, bromoform). Many halocarbons have only an anthropogenic origin (e.g., CFCs, HCFCs, HFCs) whereas others are of mixed anthropogenic and natural origin (e.g., MeBr). The molecular composition of the various CFCs, HCFCs, and HFCs can be deduced from their numerical classification by adding 90 to their designated numbers. The first digit (hundreds place) refers to the number of carbon atoms, the second to the number of hydrogens, and the third to the number of fluorines. For example, the molecular structure of CFC-12 is obtained from the resultant number of 102, and refers to a one carbon molecule containing no hydrogens and two fluorines. By inference from its CFC designation two chlorines must also be included. Similarly, the compound HFC-134 has a resultant sum of 224, which implies a two carbon molecule containing two hydrogens and four fluorines (tetrafluoroethane). Some halocarbons, such as CFCs, HCFCs, and HFCs, absorb infrared energy reflected from the earth's surface and are therefore referred to as "greenhouse" or "radiatively-active" gases. All halocarbons of this type undergo photodissociation reactions in the upper atmosphere whereby chlorine and bromine atoms are liberated and contribute

NATO ASI Series, Vol. I 39
Microbiology of Atmospheric Trace Gases
Edited by J. Colin Murrell and Donovan P. Kelly

to the destruction of stratospheric ozone as was first noted by Molina and Rowland (1974). This process is illustrated in Fig. 1. These ozone-degrading reactions may be summarized as follows:

$$Cl + O_3 \longrightarrow ClO + O_2 \quad (1)$$

$$Br + O_3 \longrightarrow BrO + O_2 \quad (2)$$

$$ClO + BrO \longrightarrow Cl + Br + O_2 \quad (3)$$

Reaction (2) is perhaps as much as 100-fold more efficient than reaction (1) in scavenging ozone, and hence the upward flux of bromine-containing halocarbons to the stratosphere is a matter of great environmental concern. Reaction (3) regenerates the reactive halogen atoms, and may account for as much as 20 to 30 % of current stratospheric ozone depletion, as well as for the ozone loss over polar regions (Anderson et al, 1991; Toohey et al, 1990; Wahner and Schiller, 1992). Because of the threat these substances pose to the stratospheric ozone layer, international agreements (e.g., The Montreal Protocol) have been made to ban or constrain the use of certain halocarbons and to seek environmentally-acceptable substitutes (Fischer et al., 1991; MacFarland, 1992).

An important factor to be considered in understanding the cycling of halomethanes in the atmosphere is their residence times, which are calculated from the experimentally-determined kinetics of their destruction via oxidation with hydroxyl radicals. In general, the longer the residence time of a given halocarbon in the troposphere, the greater will be the quantity that penetrates the tropopause and enters the stratosphere where it will contribute to ozone decay. The annual flux of a given halomethane from the earth's surface to the atmosphere can be calculated by dividing the amount of the substance present in the atmosphere as determined from global sampling programs (referred to as the "atmospheric burden") by the residence time. For comparative purposes, the atmospheric residence time for methane is 8 to 12 years (Cicerone and Oremland, 1988) whereas those for the chemically-unreactive CFCs range from several decades to centuries in duration. HCFCs and HFCs, which were designed as replacements for CFCs, have residence times of about 15 years, whereas the highly reactive MeBr has a residence time of about 2 years (Khalil et al., 1993; Singh and Kanakidou, 1993) and perhaps as short as 1.2 years if the oceans constitute a net sink for MeBr (Lobert et al., 1995). Because calculation of residence times is based only on atmospheric chemistry, it ignores the possibility of microbial consumption from the atmosphere by soil and aquatic microbes. In the case of methane, consumption by soil methanotrophs may account for between 1 and 10 % of the annual methane budget (Born et al., 1990; Crutzen, 1991). The remainder of this paper will discuss what is known about microbial degradation of halocarbons which are present in the atmosphere.

Chlorofluorocarbons (CFCs)

CFCs are unreactive and hence are useful as refrigerants and propellants. Because of their recalcitrance to chemical and/or biological degradation, they have achieved a ubiquitous distribution in the biosphere, and are commonly employed as tracers of water movement and atmospheric ventillation or dating of ocean and ground waters (Bullister and Weiss, 1983; Fine and Molinari, 1988; Busenberg and Plummer, 1992). Nonetheless, bacterial degradation of CFCs has been observed. Removal of CFC-11 and CFC-12 at atmospheric mixing ratio levels was reported from flux chambers emplaced over termite mounds and rice paddies (Khalil and Rasmussen, 1989; Khalil et al., 1990). Semprini et al. (1992) noted the disappearance of both CFC-11 and CFC-113 in groundwater systems undergoing active denitrification. Lovley and Woodward (1992) carried out mechanistic studies and demonstrated that both CFC-11 and CFC-12 were consumed during anoxic incubations of sediments. The investigators sealed low levels of the CFCs (100 to 1000 ppt) into bottles and were able to observe their disappearance to below atmospheric mixing ratios. Degradation of CFC-12 was clearly achieved by biological activity, but CFC-11 may have been degraded by a combination of biological and chemical processes. The products of the consumption reactions could not be determined because analyses were performed on an electron-capture gas chromatograph which becomes selectively less sensitive to products as halogens are removed from the carbon atom. Krone and Thauer (1992) noted consumption of CFC-11 by *Methanosarcina barkerii*. Because these workers were employing very high concentrations of CFC-11 (~1,600 ppm) they were able to detect HCFC-21 and HCFC-31 as products indicating that reductive dechlorination reactions occurred. However,

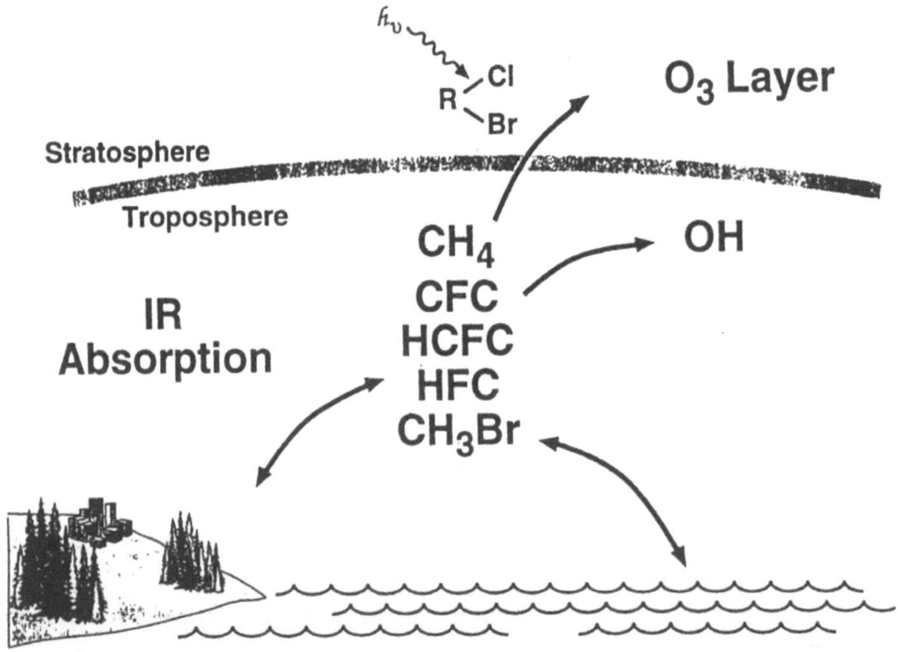

Figure 1: A schematic representation of the sources, sinks, and environmental perturbations caused by atmospheric halocarbons and methane. The doubly-pointed arrows indicate that the locations can be both sources and sinks for the compound in question.

because carbon monoxide and fluoride were also found, a complete mineralization may have ultimately occurred. Methanotrophs do not appear to be able to degrade CFC-11 (DeFlaun et al., 1992) or CFC-12 (Oremland, RS, unpublished data). The global significance of the anaerobic mineralization of CFCs is uncertain. Clearly, because they are so readily detected in the environment and usually appear at concentrations which indicate no consumptive reactions occur, the biological cycling of CFCs appears to be a minor sink term. A crude estimate compared to known sinks (atmospheric hydroxyl radicals) suggests that biodegradation may account for the removal of as much as 5 % of CFC-12 and 15 % of CFC-11 (Khalil et al., 1990).

Hydrochlorofluorocarbons (HCFCS) and Hydrofluorocarbons (HFCS)

Methanotrophs co-metabolize a number of chlorinated pollutant halocarbons (McCarty and Reinhard, 1993), and the possibility exists that this process could extend to HCFCs. *Methylosinus trichosporium* was screened for its ability to degrade several halocarbons (DeFlaun et al., 1992). At high mixing ratios (~ 2,750 ppm) consumption of HCFCs 21, 141b, 131, and HFC 143 occurred, but not of HCFCs 123, 142b, or of HFCs 134a or 134. A stoichiometric recovery of Cl^- and F^- in the spent medium was noted which implied complete mineralization of the halocarbons to carbon dioxide. MeF, which inhibits methane and ammonia monooxygenases (Oremland and Culbertson, 1992 a & b; Miller et al., 1993), is oxidized by *Methylococcus capsulatus* when present at less than fully-inhibitory levels. Methane slowed the rate of MeF consumption which suggests action as a competitive inhibitor (Fig 2). Consumption of HCFCs 21 and 22 was observed for cell suspensions of *M. capsulatus* as well as for soil (Table 1). In general, HCFC-21 was more readily degraded than HCFC-22, and the presence of methane was required in order for the HCFC to be consumed. Degradation of the

HCFC	Conc'n	System	% Removal	Time (d)
HCFC-21	100 ppm	M.capsul.	98	2.0
HCFC-21	2-10 ppm	M.capsul.	64-94	5.5
HCFC-21	100 ppm	Soil	100	19.0
HCFC-22	1000 ppm	M.capsul.	57	2.0
HCFC-22	1000 ppm	Soil	0	32.0
HCFC-22	25 ppm	Soil	68	47.0

Table 1: Consumption of HCFC-21 and HCFC-22 by *M. capsulatus* and soils.

HCFCs was far more rapid in cell suspensions than in soils. Indeed, at some of the high concentrations(1000 ppm) biodegradation of HCFC-22 was not observed in soil, and only by lowering the mixing ratio to 25 ppm did a partial consumption occur over prolonged incubation. It appears that HCFCs are also competitive inhibitors of methane monooxygenases, as is MeF. No HCFC consumption occurred in a variety of controls (autoclaved, anoxic, or lacking methane). These results indicated that methanotrophs in natural assemblages could degrade HCFCs 21 and 22. However, although testing dense cell cultures for their capacity to metabolize HCFCs has its merit, it is also necessary to conduct similar experiments with natural populations to see if the results can be extended to field studies. Furthermore, although these types of results qualitatively indicate that consumption of certain HFCs and HCFCs from the atmosphere is at least theoretically possible, experiments need to be conducted with realistic mixing ratios (low ppt levels) to determine if microbes can attack HCFCs and HFCs under field situations. Consumption of trace levels (ppb) of HCFC-21, HCFC-141b, and HCFC-123 has been observed during incubation of anoxic sediments (D. Lovley and D. Lonnegran, unpublished data).

Methyl Halides

Methyl halides of interest to atmospheric chemistry include methyl chloride (MeCl), methyl bromide (MeBr), and methyl iodide (MeI). In aqueous solution these substances can undergo a variety of nucleophilic substitution reactions with other halogen ions, OH^- or HS^- to yield other methyl halides, methanol, or methylated sulfur compounds like methanethiol (MeSH) and dimethylsulfide (DMS) (Swain and Scott, 1953; Zafiriou, 1975; Elliott and Rowland, 1993; Oremland et al, 1994 a). MeBr is used extensively as a fumigant to eliminate pests, and has come into increased usage in agriculture to pre-treat fields before planting with various fruits (e.g., strawberries) and vegetables. The global atmospheric burden of MeBr has been estimated to be about 200 x 10^6 Kg, and because it has a residence time of about 2 years, an annual flux to the atmosphere from the earth's surface is calculated to be about 100 x 10^6 kg (Khalil et al., 1993; Singh and Kanakidou, 1993). About one third of the annual flux to the atmosphere is thought to arise from anthropogenic sources. Because of the high solubility of MeBr in water (KH = about 0.24), the oceans play a critical role in buffering the MeBr mixing ratio of the atmosphere (Butler, 1994). Indeed, the oceans appear to be a net sink rather than a source for atmospheric MeBr (Lobert et al., 1995). This fact decreases its estimated atmospheric residence time to 1.2 years. However, if the 2 year residence time is still realistic, then the annual outward flux from the terrestrial environment must have been underestimated. A variety of aerobic and anaerobic bacteria can degrade methylated halides. Under anaerobic conditions, MeCl undergoes a biologically-catalysed reaction to form MeSH (Braus-Stomeyer et al., 1993), and MeCl has been shown to support the growth of acetogenic bacteria (Traunecker et al., 1991;). In anoxic environments having free HS^-, MeBr will chemically react to form MeSH and DMS which are subsequently attacked by

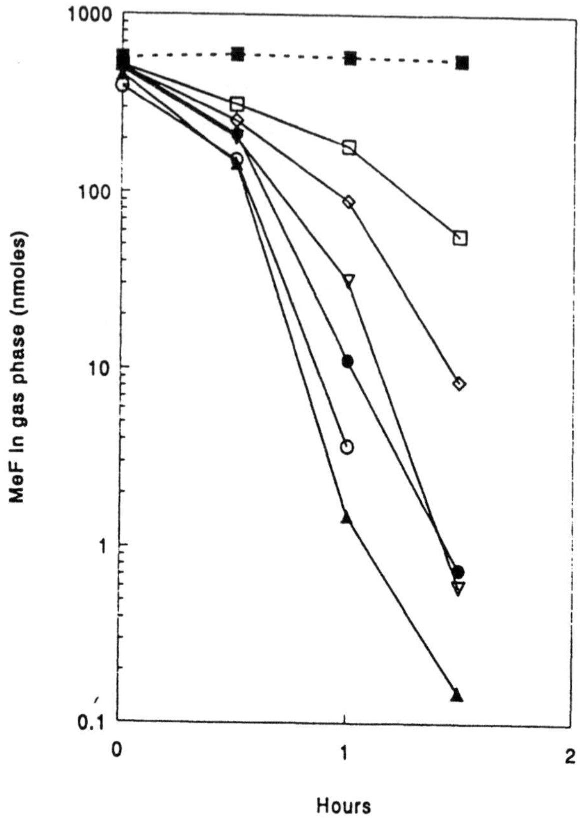

Figure 2: Consumption of MeF by cell suspensions of M. capsulatus incubated under the following percent levels of CH4 in the headspace: 2.5 (o), 5 (Δ), 10 (o), 15 (∇), 25 (◊), 40 (□), and a heat-killed control (■). Twenty ml of cell suspension were incubated under a 37 ml gas phase.

methanogenic and/or sulfate-reducing bacteria (Oremland et al., 1994 a). Experiments with ^{14}C-MeI confirmed that $^{14}CH_4$ plus $^{14}CO_2$ were the products under methanogenic conditions whereas $^{14}CO_2$ was the sole product under sulfate-reducing conditions.

Under aerobic conditions, MeBr acts as a methane analogue and MeBr is oxidized by cell suspensions of M. capsulatus. However, high MeBr levels (10,000 ppm) inhibit methane oxidation in soils (Oremland et al., 1994 b), lower levels of added MeBr (< 1,000 ppm) could be consumed by soil methanotrophs. Experiments with ^{14}C-MeBr confirm its oxidation by soil methanotrophs and/or nitrifiers to $^{14}CO_2$. However, aerobes other than methanotrophs and nitrifiers also appear to be able to consume 10 ppm MeBr since soils incubated with the specific inhibitor methyl fluoride also consumed MeBr, but at a slower rate than was observed without the inhibitor (Oremland et al., 1994 b). Consumption of trace levels (20 ppm) of MeBr

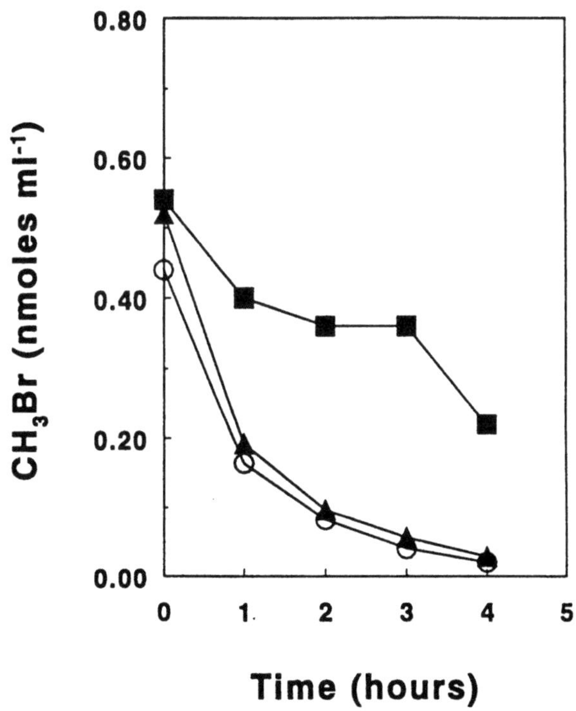

Figure 3: Consumption of 20 ppm MeBr by moist methanotrophic soils. Symbols: o, anoxic; ▲, aerobic; ■, autoclaved.

occurs rapidly, and under aerobic or anaerobic conditions rates can be nearly equivalent (Fig 3).

Field studies of MeBr flux from fumigated fields have indicated by mass balances that most (about 90 %) of the applied MeBr escapes to the atmosphere (Yagi et al., 1993). However, in more alkaline soils recovery of Br- as well as flux measurements indicated that about 70 % of the added MeBr was mineralized whereas about 30 % escaped to the atmosphere (Yagi et al., 1995). These results suggest that soil factors such as moisture and pH can greatly affect mineralization by biological/chemical processes and hence affect the amount released to the atmosphere. Experiments made with ^{14}C-MeBr during fumigation events indicate that about 10 % of the MeBr injected into strawberry fields is oxidized to $^{14}CO_2$ (L. Miller, T. Connell, and R. Oremland, unpublished data). Rapid oxidation of MeBr by soil microbes therefore
occurs during field fumigation events when MeBr levels in the soil pore spaces are very high as well as when MeBr is present at trace (low ppm) levels.

The question remains whether soil microbes can consume MeBr from the atmosphere. Recently, Shorter et al. (1995) reported uptake of near-ambient mixing ratios of MeBr (~ 10 ppt) in field-incubated flux chambers, and that MeBr levels declined to sub-ambient during the incubation. The activity was inhibited by autoclaving as well as by selected biocides (P. Crill, pers. comm.) which indicated a microbial origin. Microbial consumption by soils could further decrease the estimated atmospheric residence time of MeBr to < 0.9 years (Shorter et al., 1995). Uptake by natural microbial systems at near-in situ levels of MeBr in the oceans (e.g., picomolar) has also been observed. Unfiltered tropical ocean water samples removed MeBr about 40 % more rapidly than filtered samples over a time course of 150 h (E. Saltzman, pers. comm.).

Trifluoroacetate (TFA)

The discontinuation of CFCs and HCFCs usage called for in the Montreal Protocol has prompted an intense search for environmentally acceptable alternatives. Proposed replacement compounds include HFC 134a, HCFC 123, and HFC 143a (Wallington et al., 1994). In the atmosphere, these replacement molecules are predicted to react with hydroxyl radicals and yield a number of intermediates of varying longevity, depending upon the presence or absence of other atmospheric constituents (e.g., NO_2). One of the key major end-products is trifluoroacetylfluoride, which when encountering water droplets present in cloud formations will hydrolyze to trifluoroacetate (Franklin, 1993; Wallington et al., 1994). Figure 4 illustrates the predicted atmospheric pattern of degradation for HFC 134a to TFA. The TFA concentration predicted to occur in rainwater is about 1 nM (Franklin, 1993; Mogelberg et al., 1994). Concern exists with regard to the final fate of TFA in the terrestrial and marine environments.

In certain evaporative environments, such as closed basin lakes and vernal pools, models predict that the concentration of TFA may rise to micromolar levels within a decade or two (Tromp et al., 1995). Because TFA is a perfluorinated compound and is therefore analogous to teflon, it is extremely resistant to chemical and/or biological degradation. Although there are a number of examples of microbial defluorination reactions for monofluorinated aliphatic and aromatic compounds (Fetzner and Lingens, 1994), the only report of defluorination of compounds with multiple attached fluorines was the work of DeFlaun et al. (1992) with methanotrophs. The possibility that TFA could undergo biodegradation was raised by Visscher et al. (1994), who reported that under methanogenic conditions 2-^{14}C-TFA was quantitatively converted to $^{14}CH_4$ in saltmarsh and freshwater lake sediments. The degradation of TFA was achieved by a sequence of reductive defluorinations which resulted in the transient appearance of 2-^{14}C-labelled difluoroacetate, monofluoroacetate, and acetate as intermediates. Incubation under sulfate-reducing conditions resulted in the production of $^{14}CO_2$ in lieu of $^{14}CH_4$

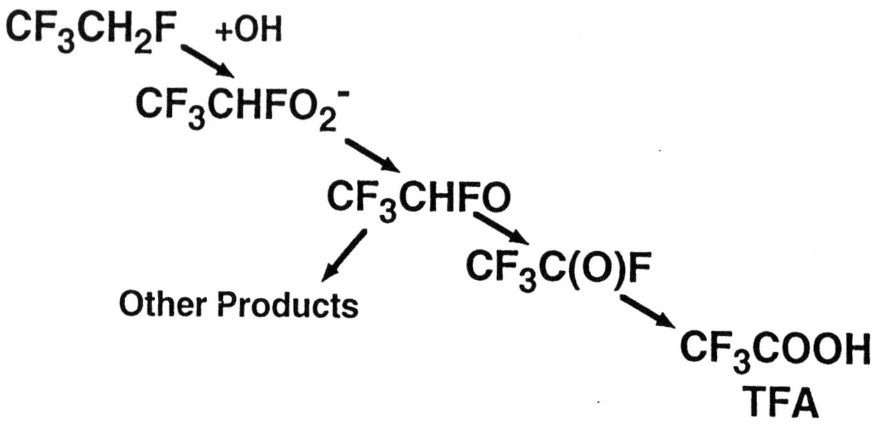

Figure 4: Atmospheric degradation pathway giving rise to TFA predicted for HFC 134a.

monofluoroacetate gave some concern because it is a potent TCA cycle inhibitor. No degradation was observed under denitrifying conditions. Under aerobic incubation, the authors reported ^{14}C-labelled fluoroform (HCF_3) as the product. This observation raised additional concerns because fluoroform is a long-lived stratospheric gas which has "greenhouse" warming potential.

The products of microbial TFA degradation are summarized in Figure 5. Subsequent experiments with these sediments have repeated these observations; however, for some as-yet-unknown reason, biodegradation of TFA could not always and easily be consistently induced in these sediments, and biodegradation of TFA has not been confirmed by other laboratories (C. Criddle, M. Odum, pers. comm.). Hence, TFA must be viewed as a highly refractory compound which nonetheless appears to be susceptible to biodegradation under certain conditions. It remains to be determined whether this observed TFA degradation phenomenon will influence the predicted patterns of global TFA accumulation and whether significant biomineralization of TFA will occur in nature.

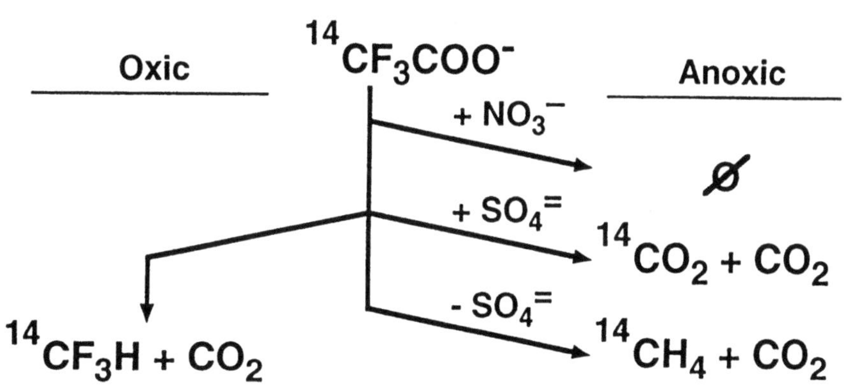

Figure 5: A summary of the endproducts of biodegradation of 2-14C-TFA reported by Visscher et al. (1994) during anoxic incubation of sediments.

Conclusions

Experiments with both pure cultures and sediment/soil systems have now shown that biodegradation of many representative forms of halocarbons (CFCs, HCFCs, HFCs, methyl halides) and their derivatives (i.e., TFA) is theoretically possible. It remains to be determined, however, if these reactions have a significant influence on the atmospheric residence times of these gases, or upon the mineralization of TFA in the biosphere.

Acknowledgements

I am grateful to K. Kvenvolden and L. Matheson for their constructive reviews of this manuscript. This paper is contributed in memory of my father, Murray Oremland.

References

Anderson DL, Toohey DW, Brune WH (1991) Free radicals within the Antarctic vortex: The role of CFCs in Antarctic ozone loss. Science 251: 39 - 46

Born M, Dorr H, Levin I (1990) Methane consumption in aerated soils of the temperate zone. Tellus 42B: 2 - 8

Braus-Stomeyer S, Cook AM, Leiseinger T (1993) Biotransformation of chloromethane to methanethiol. Environ Sci Technol 27: 1577 - 1579

Bullister JL, Weiss RF (1983) Anthropogenic chlorofluoromethanes in the Greenland and Norwegian Seas. Science 221: 265 - 268

Busenberg E, Plummer N (1992) Use of chlorofluorocarbons (CCl_3F and CCl_2F_2) as hydrologic tracers and age dating tools: The alluvium and terrace system of Central Oklahoma. Water Resources Res 28: 2257 - 2284

Butler JH (1994) The potential role of the oceans in regulating atmospheric CH3Br. Geophys. Res. Lett. 21:185 - 188

Cicerone RJ, Oremland RS (1988) Biogeochemical aspects of atmospheric methane. Global Biogeochem. Cycles 2: 299 - 327

Crutzen PJ (1991) Methane's sinks and sources. Nature 350:380 - 381

DeFlaun MF Ensley BD, Steffan RJ (1992) Biological oxidation of hydrochlorofluorocarbons (HCFCs) by a methanotrophic bacterium. Bio/Technology 10: 1576 - 1578

Elliott S, Rowland FS (1993) Nucleophilic substitution rates and solubilities for methyl halides in seawater. Geophys Res Lett 20: 1043 - 1046

Fetzner S, Lingens F (1994) Bacterial dehalogenases: biochemistry, genetics, and biotechnological implications Microbiol Rev 58: 641 - 685

Fine RA, Molinari RL (1988) A continuous deep western boundary current between Abaco (26.5°N) and Barbados (13°N). Deep - Sea Res 35: 1441 - 1450

Fischer, SK, Huges, PJ, Fairchild, PD, Kusik, CL, Dieckmann, JT, McMohn, EM, and Hobday, N (1991) Energy and global warming impacts of CFC alternative technologies. Executive Summary sponsored by AFEAS, Wash, DC, and US Dept Energy, Oak Ridge, TN, 23 p

Franklin J (1993) The atmospheric degradation and impact of 1,1,1,2-tetrafluoroethane (hydrofluorocarbon 134a). Chemosphere 27: 1565 - 601

Kahlil MAK, Rasmussen RA (1989) The potential of soils as a sink of chlorofluorocarbons and other man-made chlorocarbons. Geophys Res Lett 16: 679 - 682

Khalil MAK, Rasmussen RA, Wang MX, Ren L (1990) Emissions of trace gases from Chinese rice fields and biogas generators: CH_4, N_2O, CO, CO_2, chlorocarbons and hydrocarbons. Chemosphere 20: 207 - 225

Khalil MAK, Rasmussen RA, Gunawardena RJ (1993) Atmospheric methyl bromide: trends and global mass balance. Geophys Res Lett 98: 2887 - 2896

Krone UE, Thauer RK (1992) Dehalogenation of trichlorofluoromethane (CFC-11) by *Methanosarcina barkerii*. FEMS Microbiol Lett 90: 201 - 204

Lovley DR, Woodward JC (1992) Consumption of freons CFC-11 and CFC-12 by anaerobic sediments and soils. Environ Sci Technol 26: 925 - 929

Lobert JM, Butler JH, Montzka SA, Geller LS, Myers RC, Elkins JW (1995) A net sink for atmospheric CH3Br in the east Pacific Ocean. Science 267: 1002 - 1005

McCarty PL, Reinhard M (1993) Biological and chemical transformations of halogenated aliphatic compounds in aquatic and terrestrial environments. p 839 - 852 in Oremland, RS (ed) Biogeochemistry of global change: Radiatively active trace gases, Chapman and Hall, NY.

MacFarland M (1992) Investigations of the environmental acceptability of fluorocarbon alternatives to chlorofluorocarbons. Proc Nat'l Acad Sci. USA 89: 807 - 811.

Miller LG, Coutlakis MD, Oremland RS, Ward BB (1993) Selective inhibition of ammonium oxidation and nitrification-linked N_2O formation by methyl fluoride and dimethyl ether. Appl Environ Microbiol 59: 2457 - 2464

Mogelberg TE, Nielsen OJ, Sehested J, Wallington TJ, Hurley MD (1994) Atmospheric chemistry of CF_3COOH. Kinetics of the reaction with OH radicals. Chem Physics Lett 226: 171 - 177
Molina M, Rowland FS (1974) Stratospheric sink for chlorofluoromethanes: Chlorine atom catalysed destruction of ozone. Nature 249: 810 - 812
Oremland RS, Culbertson CW (1992a) Evaluation of methyl fluoride and dimethyl ether as inhibitors of aerobic methane oxidation. Appl Environ. Microbiol. 58: 2983 - 2992
Oremland RS, Culbertson, CW (1992b) Importance of methane-oxidizing bacteria in the methane budget as revealed by the use of a specific inhibitor. Nature 356: 421 - 423
Oremland, RS, Miller, LG, and Strohmaier, FE (1994 a) Degradation of methyl bromide in anaerobic sediments. Environ Sci Technol 28: 514 - 520
Oremland RS, Miller LG, Culbertson CW, Connell TL, Jahnke L (1994b) Degradation of methyl bromide by methanotrophic bacteria in cell susspensions and soils. Appl Environ Microbiol 60: 3640 - 3646
Semprini L, Hopkins GD, Roberts PV, McCarty PL (1992) In-situ transformation of carbon tetrachloride and other halogenated compounds resulting from biostimulation under anoxic conditions. Environ Sci Technol 26: 2454 - 2461
Shorter JH, Kolb CE, Crill PM, Kerwin RA, Talbot RW, Hines ME, Harriss RC (1995) An effective soil surface sink for atmospheric methyl bromide. (ms in review)
Singh HB, Kanakidou M (1993) An investigation of the atmospheric sources and sinks of methyl bromide. Geophys Res Lett 20: 133 - 136
Swain CG, CB Scott (1953) Quantitative correlation of relative rates. Comparison of hydroxide ion with other nucleophilic reagents towards alkyl halides, esters, epoxides, and acyl halides. J Amer Chem Soc 75: 141 - 147
Toohey DW, Anderson JG, Brune WH, Chan KR (1990) In situ measurements of BrO in the Arctic stratosphere. Geophys Res Lett 17: 513 - 516
Traunecker J, Preu, A, Diekert G (1991) Isolation and characterization of a methyl chloride utilizing, strictly anaerobic bacterium. Arch. Microbiol. 156: 416 - 421
Tromp TK, Ko MKW, Rodriguez, Sze ND (1995) Alternative CFCs: Accumulation of trifluoroacetic acid in seasonal wetlands. (manuscript in review)
Visscher PV, Culbertson CW, Oremland RS (1994) Degradation of trifluoroacetate in oxic and anoxic sediments. Nature 369: 729 - 731
Wahner A, Schiller A (1992) Twilight variation of vertical column abundances of OClO and BrO in the north polar region. J Geophys Res 97: 8047 - 8055
Wallington TJ, Schneider WF, Worsnop DR, Nielsen OJ, Sehested J, Debruyn WJ, Shorter, JA (1994) The environmental impact of CFC replacements - HFCs and HCFCs. Environ Sci

impact of CFC replacements - HFCs and HCFCs. Environ Sci Technol 28: 320 - 326
Yagi, K, Williams, NY, Cicerone, RJ (1993) Agricultural soil fumigation as a source of atmospheric methyl bromide. Proc Nat'l Acad Sci USA 90: 8420 - 8423
Yagi K, Williams J, Wang N-Y, Cicerone RJ (1995) Atmospheric methyl bromide (CH_3Br) from agricultural soil fumigations. Science 267: 1979 - 1981
Zafiriou OC (1975) Reaction of methyl halides with seawater and marine aerosols. J Mar Res 33: 75 - 81

Microbial Transformation of Alkyl Esters

Donovan P Kelly and Ann P Wood[1]

Institute of Education, University of Warwick, Coventry CV4 7AL, England, and [1]Division of Life Sciences, King's College London, Campden Hill Road, London W8 7AH, England

Introduction

Low molecular weight organic acids and alcohols, such as formate, acetate, methanol and ethanol are well known as growth substrates for bacteria, and can be metabolized by autotrophic (formate), methylotrophic (formate, methanol) or heterotrophic pathways (acetate, ethanol). Less is known about the biodegradation of the esters such as methyl formate, ethyl acetate and so on, up the aliphatic series of possible combinations, although the biological origin of some such esters, such as ethyl acetate is well established (Armstrong et al. 1984; Longo et al. 1992; Kallelmiri and Miclo 1993), and some more complex esters have been demonstrated to be animal pheromones (Mori and Puapoomchareon 1990). Numerous biogenic and anthropogenic sources obviously exist for alcohols, organic acids and their esters, including ripening fruits (Mattheis et al. 1991), alcoholic and acid fermentation reactions (both bacterially- and yeast-catalyzed), in the urban atmosphere and during smog aerosol events (Grosjean 1992; Schultztokos et al. 1992), and numerous organic acids, alcohols and esters, including ethyl acetate, and ketones, such as acetone, are secreted by animals, including dogs, foxes, hyenas and lions (Albone 1984; Natynczuk et al. 1989; Darling 1990), in human breath (Manolis 1983) and by plants (Whitfield et al. 1981). Numerous examples of acids and esters are found in the anal sac secretions of some animals, and the mix of volatile chemicals is postulated to arise through microbial catalysis within these specialized organs (Albone and Perry 1976; Albone et al. 1976; Albone 1984). The scores of compounds identifiable in the

NATO ASI Series, Vol. I 39
Microbiology of Atmospheric Trace Gases
Edited by J. Colin Murrell and Donovan P. Kelly

breath of normal healthy human beings include ethanol, ethyl acetate and acetone at exhalation rates averaging 50-500 μg/h (Conkle et al. 1975). When extrapolated to a world population of five billion, this indicates a combined output of these compounds of around thirty thousand tonnes per year. Added to industrial sources, microbial fermentations, and the exhalations of other mammal and non-mammalian animals, the input of low molecular weight esters into the atmosphere is likely to constitute a significant trace component of the global carbon cycle. Formic and acetic acids are apparently generated in the atmosphere by photochemical processes, and probably have an important role in the global cycling of carbon, particularly as carbon inputs to the ocean (Morikami et al. 1993).

Previous work on ester metabolism by bacteria has included the demonstration that the use of methyl formate by methanotrophs is apparently dependent on its spontaneous hydrolysis to give methanol, which is a growth substrate for methylotrophs (Meyers 1982). Growth of the newly isolated *Bacillus thermoglucosidasius* on ethyl acetate was apparently dependent on an esterase, excreted into the medium only late in growth, and subject to regulation by catabolite repression (Stalder and Bachofen 1994). *Pseudomonas fluorescens* produces intracellular esterase, and hydrolyses methyl butyrate with a lipase (Nakagawa et al. 1984a,b), while acetone has been shown to be degraded by aerobes, some sulfate-reducing bacteria, denitrifying bacteria and non-sulfur purple phototrophic bacteria (Taylor et al. 1980; Platen and Schink 1989; Platen et al. 1990; Madigan 1990).

We report here some comparative observations on the metabolism of some low molecular weight organic acids, alcohols and their esters, and acetone, by growing cultures of the metabolically versatile facultative chemolithotroph and methylotroph, *Paracoccus versutus* (Katayama et al. 1995).

Organisms and procedures

Paracoccus versutus (formerly known as *Thiobacillus versutus*; Kelly and Harrison 1989) was used as the test organism. The wild type (A2) and a strain capable of constitutive fast growth on glucose (GF1; Wood and Kelly 1978), were maintained on agar slants of media with methylammonium chloride (50 mM), sucrose (10 mM) or glucose (20 mM, strain GF1 only), and these used to inoculate liquid medium cultures containing diverse organic substrates, essentially as described previously (Wood and Kelly 1977, 1983a,b; Kelly and Wood 1982, 1984; Kelly et al. 1979). Cultures were grown at 30°C, and liquid cultures (25 ml) were shaken in 250 ml Erlenmeyer flasks in reciprocating shakers (LH Engineering or Gallenkamp). All cultures provided with volatile substrates were grown in flasks sealed with Suba-seal® vaccine stoppers. Growth was determined either by direct measurement of increase in turbidity using flasks with Pyrex® glass side-arms which could be inserted directly into a colorimeter, or by removal of samples (1 ml) for measurement of OD (1 cm cuvette; 440 nm) in a spectrophotometer. Optical density (OD) readings were converted to biomass by means of dry weight calibration curves prepared for the two strains.

Results and Discussion

Growth on methyl compounds, ethanol and acetone

Extensive work has been done on the use of formate, formaldehyde, methanol and methylammonium by *P. versutus* (Kelly and Wood 1982, 1984; Kelly et al. 1979; Wood and Kelly 1981, 1983a; Kraczkiewicz-Dowjat and Kelly 1991). We have also confirmed that the *P. versutus* strains are unable to grow on dimethyl sulfide, dimethyl sulfate, formamide, dimethyl formamide, nitromethane or carbon monoxide. Formamide was, however, not toxic even at 100 mM (allowing growth to occur on 50 mM methylammonium chloride in its presence). Growth rates on single C_1- and C_2-compounds by both strains are summarized in Table 1.

Approximate growth yields for some compounds of interest were determined from the biomass produced in batch cultures as follows (g dry weight/mol substrate): formate 4.2 ± 1.3 [8 values]; acetate 19.8 ± 2.3 [3]; methanol 5.5 ± 0.8 [3] (Kelly and Wood 1982); ethanol 25.7 ± 4.0 [7]; dimethyl ketone (acetone) 23.7 ± 2.1 [3]; methylammonium (MMA) 9.5 ± 0.7 [3]. The yield on formate was similar to that determined previously by different means (Kelly et al. 1979), but that for methylammonium under these growth conditions was lower than that

Table 1. Growth rates of *Paracoccus versutus* strains on C_1- and C_2-compounds. Early stationary phase cultures on 20 mM acetate (wild type A2), 10 mM succinate (A2), 50 mM methylammonium (MMA; strain GF1) or 20 mM glucose (GF1) were inoculated into flasks containing the indicated single substrates and growth measured for 20-240 h, depending upon the growth rates expressed. All rates are means of several (2-6) determinations, and are expressed as doubling times (T_d, hours) and specific growth rate (μ, h^{-1}).

Substrate	(mM)	Wild type		Strain GF1	
		T_d	μ	T_d	μ
Formate	(5)	11.9	0.058	-	-
	(10)	17.6	0.039	-	-
	(20)	-	-	[very slow]	
Acetate	(5)	2.6	0.267	-	-
	(10)	1.8	0.385	2.7	0.257
	(20)	-	-	2.1*	0.330
Methanol	(20)	27.7	0.025	222	0.003
Ethanol	(10)	2.0	0.350	1.9	0.365
Acetone	(50)	13.0**	0.053	11.8†	0.059
MMA	(20)	-	-	3.7	0.187
	(50)	3.0	0.231	4.4	0.158

* glucose inoculum; ** succinate inoculum; † acetone inoculum

in optimized batch and chemostat cultures (13 g/mol; Kelly and Wood 1984). The mechanism of acetone metabolism has not been examined further, but the growth yield (cf. acetate, methanol) indicates its energetic equivalence to acetate and methanol, which would be consistent with the hypothetical possibility of its metabolism *via* methyl acetate as an intermediate (Taylor et al. 1980). There is, however, no evidence for such a pathway and metabolism via acetol (hydroxyacetone) might be more likely (Lukins and Foster 1963; Taylor et al. 1980). Alternatively, its use might involve isopropyl alcohol (Price and Rittenberg 1950) or acetoacetyl CoA (Platen et al. 1990).

Growth on methyl formate

P. versutus strain GF1 showed a short period (9-10 h) of rapid growth (T_d, 7.5 h) when inoculated from a 50 mM MMA culture into 20 mM methyl formate medium, followed by more than 200 h of growth with a T_d of 269 h. A parallel culture (from the same inoculum) on 20 mM methanol grew with a T_d of 238 h with an increase in biomass of about 98 mg dry wt/liter (yield = 4.9 g/mol). The initial rapid phase on methyl formate accounted for about 30% of the biomass produced after a total of 255 h incubation. The increase in biomass was about 129 mg/l, indicating a gross yield of 6.5 g dry wt/mol methyl formate. This would be consistent with the bacteria deriving energy for growth from the oxidation of both methanol and formate released from the methyl formate.

When inoculated into 50 mM methyl formate the initial phase of more rapid growth (T_d, 8.3 h) was extended to 24 h. Subsequently growth continued with a T_d of about 40 h (μ = 0.177 h^{-1}). Further additions of 50 mM methyl formate after 30 and 70 h resulted in comparable 'bursts' of more rapid growth followed by steady growth with a T_d of 39-40 h. An apparent yield on methyl formate by this culture (after 234 h) was 2.8 g dry wt/mol, which is low by comparison with *P. versutus* yields on methanol alone (see above). These kinetics suggested that growth was dependent on the (chemical) hydrolysis of methyl formate (to methanol and formate) and that growth was possibly primarily at the expense of methanol.

Growth on ethyl formate

Wild type *P. versutus*, previously grown on 20 mM acetate, grew without significant lag when inoculated into media containing of one of ethanol, formate or ethyl formate (at either 5 or 10 mM). Growth on ethyl formate was rapid (T_d, 2.0-2.5 h; $\mu = 0.31\ h^{-1}$), but this ceased when the amount of growth produced was essentially identical to that in the ethanol cultures (yield about 26 g dry wt/mol), with growth continuing at about a T_d of 12.4 h in the culture with 5 mM ethyl formate, which was similar to that with formate alone (T_d, 12.0), and considerably more slowly with the 10 mM ethyl formate culture (Cf. Table 1). These results show that the wild type can immediately attack ethyl formate after growth on acetate, but grows initially on the ethanol so released, without significant simultaneous use of the formate also produced.

P. versutus strain GF1 showed biphasic growth when a culture grown on methylammonium was inoculated into ethyl formate medium (Fig. 1). The initial slower growth phase (8-10 h at T_d, 6-9 h) was essentially independent (in duration and amount of growth produced) of the ethyl formate concentration (over the range 10-40 mM), suggesting that this was a period of physiological adaptation after which the organisms were fully induced for growth on ethyl formate. The rate of the second phase of growth was essentially identical for 10, 20 and 30 mM ethyl formate (T_d, 2.45 ± 0.15 h), but may have been slightly slower with 40 mM substrate (Fig. 1). Biomass production was proportional to the ethyl formate concentration, with a mean yield of 22.9 ± 2.3 g dry wt/mol [4 determinations]. The mean yield on ethanol of 25.7 ± 4.0 g/mol indicates that only ethanol appeared to provide energy and carbon for growth during the two phases of growth on ethyl formate. Subculture into fresh medium of cultures already adapted to ethyl formate resulted in immediate growth at high rates (T_d, 2.0 and 1.8 h with 10 and 20 mM ethyl formate), although higher concentrations depressed the growth rate (T_d, 8.8 and 9.6 h with 50 and 100 mM ethyl formate). Taking an inoculum from a culture on 20 mM ethyl formate (T_d, 4.0 h) and inoculating into fresh medium with 10 or 20 mM ethyl formate produced biphasic growth curves, with T_d values of (i) 2.0 and 1.8 and (ii) 12.6 and 16.4 h respectively. Overall yields (at the times when growth had ceased) of 23.0 and 22.0 g dry wt/mol were as previously observed, but about 51 and 54% of the growth occurred in each case in the first (faster) phase. This suggests that the metabolism may not depend solely on release of ethanol and one factor could be cumulative toxicity of formate, which is known to be an inhibitory substrate (see μ in Table 1, and Kelly et al. 1979).

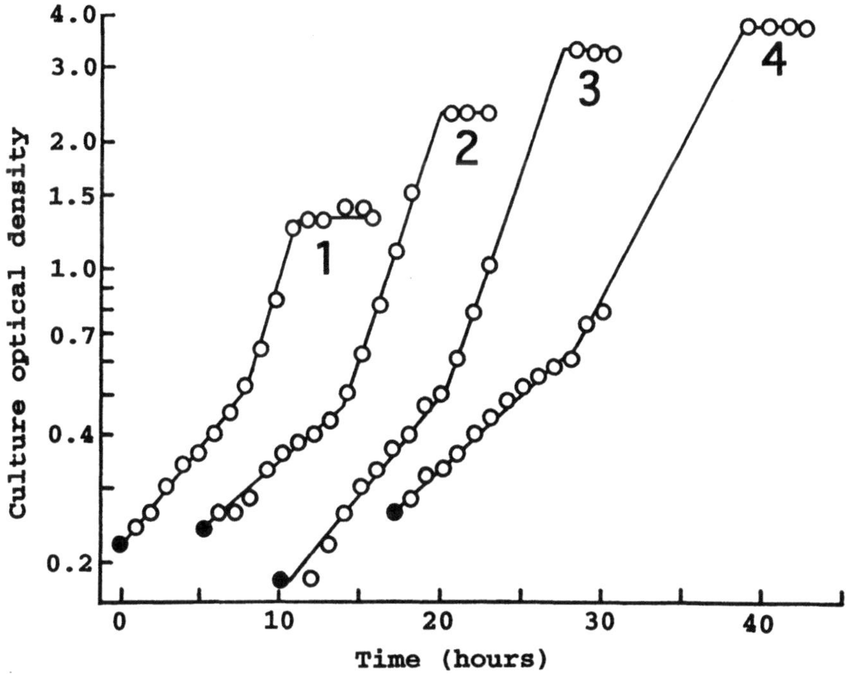

Figure 1. Growth of *Paracoccus versutus* strain GF1 on ethyl formate. Organisms previously grown autotrophically on 50 mM methylammonium chloride were inoculated into medium containing ethyl formate at 10, 20, 30 or 40 mM (curves 1-4). For greater clarity, growth curves are presented with staggering of the zero times for each culture, with the initial (inoculation time) reading for each culture being shown as a solid symbol (●). Final yields of organisms in the four cultures were 20.0, 22.0, 23.3 and 26.3 g dry wt/mol ethyl formate, with a mean of 22.9 ± 2.3 g/mol.

Growth on methyl acetate and ethyl acetate

Strain GF1 showed polyphasic growth on these substrates. Growth on methyl acetate occurred with no lag, initially with a T_d of

56 h, declining to 165 h after about 60 h growth. At 140-160 h after inoculation there was an abrupt transition to growth with T_d of 19 h. Similar results were obtained in several experiments with ethyl acetate: for example (Fig. 2) an extended period of growth at Td of 85 h abruptly switched to a T_d of around 14 h about 140 h after initial inoculation of the culture.

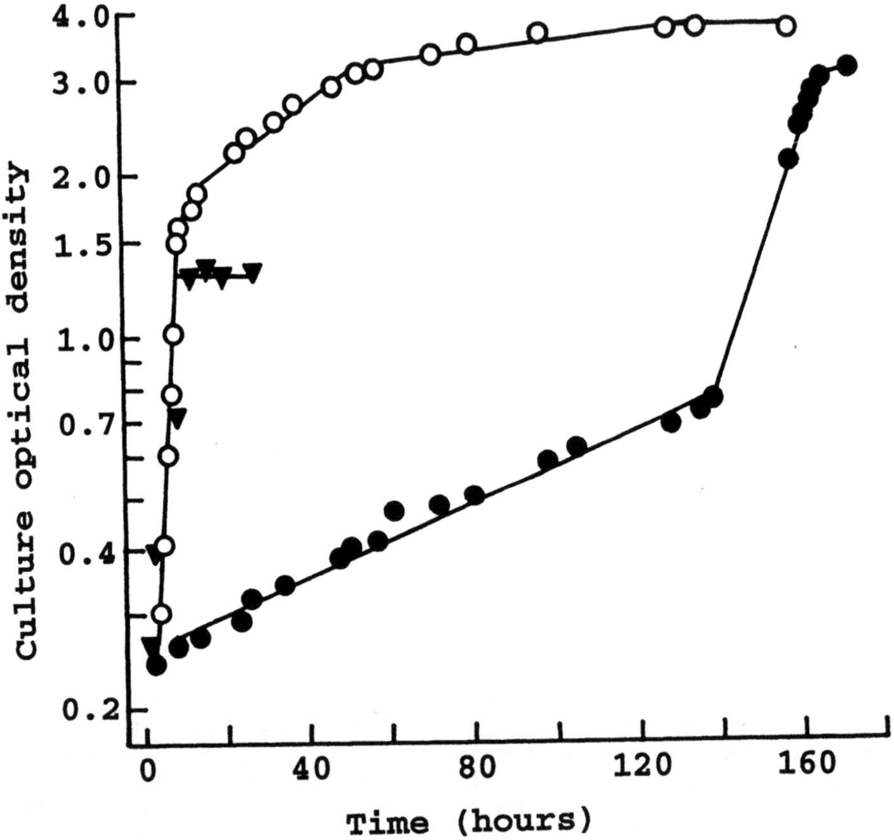

Figure 2. Growth of *Paracoccus versutus* strain GF1 on ethyl acetate, and the effect of the simultaneous presence of ethanol. Organisms previously grown on methylammonium (50 mM) were inoculated into media with 20 mM ethyl acetate (●) or 10 mM ethanol (▼) or both substrates (○). Increases in biomass (mg dry wt/liter) at cessation of growth were 613 (ethyl acetate), 252 (ethanol) and 737 (both substrates).

These results indicated adaptation to rapid growth, presumably by induction of enzymes for ester hydrolysis and acetate oxidation, only on exposure to the specific substrates. First subcultures into fresh medium with 20 mM methyl acetate or ethyl acetate also produced biphasic growth, but of a different nature: initial T_d values were faster than after sharp breaks in growth rate. For methyl acetate the two T_d phases were 6.7 and 13.1 h, and for ethyl acetate, 2.0 and 7.6 h. Biomass at the cessation of growth showed yields of 17.5 g/mol methyl acetate and 28.2 g/mol ethyl acetate, suggesting incomplete oxidation of the substrates.

Experiments were performed to see if the simultaneous presence of ethanol with ethyl acetate might enhance adaptation to use of the latter. When MMA-grown strain GF1 was inoculated into media with ethyl acetate (20 mM), ethanol (10 mM) or both substrates (Fig. 2), rapid growth occurred only with ethanol-supplemented cultures, and the kinetics of ethyl acetate use showed slow growth with a T_d of 85 h, for 140 h at which point the rate changed abruptly to a T_d of 14 h. The biomass produced indicated yields (g dry wt/mol) of 30.7 (ethyl acetate) and 25.1 (ethanol), with the biomass of 737 mg dry wt/liter in the mixed substrate culture being 85% of the sum of that for the separate substrates. The high yield on ethyl acetate indicated that some energy and carbon was derived from the acetate moiety as well as from the ethanol in this experiment.

In another test, inoculating MMA-grown strain GF1 into ethyl acetate (10 mM) alone or supplemented with ethanol (at 5 or 10 mM) produced immediate fast growth when ethanol was present (T_d, 2.9 h), but this slowed abruptly after 5 h (5 mM ethanol) or 10 h (10 mM ethanol) to new rates of (T_d) 49 and 71 h, which subsequently further decreased to T_d of 125 and 236 h. With ethyl acetate alone very slow growth (T_d, 136 h) occurred for 155 h before the abrupt adaptive transition to a T_d of 29 h. The presence of ethanol thus did not obviously aid induction of ability to use ethyl acetate. Final biomass levels (mg dry wt/l) in the three cultures (after 183-237 h) were 279 (10 mM ethyl acetate alone), 390 (+5 mM ethanol) and 560 (+10 mM ethanol). The mean growth yield found on ethanol as a sole substrate was 25.7 g/mol (see above), so the biomass due to ethanol in the +5 mM and +10 mM cultures should be 129 and 257 mg/l respectively, leaving 261 and 303 mg/l attributable to the use of ethyl acetate, which are very similar to the observed value of 279 mg/l with ethyl acetate alone. The average of these three values indicated a yield due to ethyl acetate of 28.1 g/mol. The maximum growth yield obtainable from ethyl acetate should in theory be close to the sum of yields on acetate and ethanol (but probably slightly less than this, because of the hydrolysis requirement), namely, the sum of 25.7 + 19.8 = 45.5 (with an error of around ± 4).

The comparative yields and adaptation to use of ethyl acetate was further tested in an experiment in which unadapted strain GF1 was initially grown on glucose then inoculated into media containing ethyl acetate, ethanol or acetate separately and in combination (Table 2).

Table 2. Growth rates and yields of *Paracoccus versutus* strain GF1 on ethyl acetate, ethanol and acetate in various combinations (all at 10 mM). Growth rates are given as doubling times (T_d, hours), final biomass as mg dry wt/l and yield as grams dry wt/mol substrate. Where two successive growth rates occurred the duration of the first (h) is shown in brackets.

Substrate(s)	T_d	Biomass	Yield
Ethanol	2.7	258	23.0
Acetate	2.7	177	17.7
Ethanol + Acetate	2.9	469	-
Ethyl acetate [EtAc]	(i) lag (40)	-	-
	(ii) 109 (110)	-	-
	(iii) 22	230	23.0
EtAc + Ethanol	(i) 8.7 (5)	-	-
	(ii) 2.0	420	-
EtAc + Acetate	2.4	196	-

As expected, biomass production with ethanol+acetate (469 mg/l) was the sum of the biomass produced on these substrates separately (435), showing their simultaneous use. Biomass produced on ethyl acetate was only half what could be predicted from its composition (=Σ[ethanol+acetate]), and while its use was apparently accelerated in the presence of ethanol the combined biomass production was not improved. When given ethyl acetate in the presence of acetate, it is clear that no ethyl acetate was consumed within the time scale of the experiment (25 h) as the biomass production was only that due to acetate alone (with growth ceasing 6 h after inoculation).

Following further passage through culture on 10 mM ethyl acetate it was possible to obtain cultures which grew at one single rate when subcultured into either methyl acetate or ethyl acetate and gave enhanced growth yields (Table 3). After

Table 3. Growth of *Paracoccus versutus* strain GF1 on methyl acetate or ethyl acetate. Inocula from 50 mM methylammonium cultures were transferred through 10 mM ethyl acetate then on to the indicated media. Growth rates are shown as doubling times (T_d, hours), specific growth rate (μ, h^{-1}), and yields are given as g dry wt/mol of each substrate provided. Data are from duplicate determinations in each case.

Substrate	mM	T_d	μ	Yield
Methyl acetate	10	16.5	0.042	19.0
	20	13.0	0.053	16.7
Ethyl acetate	10	7.6	0.091	34.4
	20	7.2	0.096	38.1

repeated subculture on ethyl acetate, a mean value for T_d on this substrate was 10.7 ± 3.0 h [11 determinations].

Given the separate yields on methanol, ethanol and acetate to be 5.5, 25.7 and 19.8 (see above), the values of Table 3 are about 80% of the mean yields indicated for methanol + acetate or ethanol + acetate respectively. The indication is thus that fully adapted organisms effect the total dissimilation of methyl acetate and ethyl acetate.

Wild type *P. versutus* grew without lag when inoculated (from acetate pre-culture) into media with ethanol, acetate or ethyl acetate (at either 5 or 10 mM), but growth on ethyl acetate showed only a short rapid phase (T_d, 2-3 h) during which the amount of growth was no more than 10% of that on ethanol alone. A second, slower, phase followed (T_d, 16.5 and 21.4 h for 10 and 20 mM ethyl acetate). Growth on ethanol and acetate was complete within 5-7 h, but after 19 h biomass on ethyl acetate was only 20-40% of that on ethanol or acetate alone. It thus seemed to exhibit the same requirement for prolonged adaptation as did strain GF1.

Cometabolism of ethanol and methylammonium by P. versutus

Bacteria maintained on methylammonium and inoculated into media containing ethanol were constitutively able to use ethanol

Table 4. Growth of *Paracoccus versutus* strain GF1 on ethanol and methylammonium. The effect of prior cultivation on either methylammonium (MMA) or ethanol (Et) on the subsequent ability to use these substrates alone or in combination is shown. Growth rate as doubling time (T_d, hours) and biomass production (as mg dry wt/liter) are shown. Biphasic growth (or any lag before growth) is indicated by T_d phases (i) and (ii).

Inoculum substrate	Test substrate(s) (mM)		T_d	Biomass
MMA	Ethanol (10)		2.6	251
	Methylammonium (20)		3.7	196
				(sum 447)
	Et (10) + MMA (20)		2.8	443
Ethanol	Ethanol (5)		1.7	109
	Methylammonium (20)	(i)	lag (11h)	-
		(ii)	5.9	211
	Et (5) + MMA (10)	(i)	3.0 (4h)	104
		(ii)	9.5	117
				(total 221)
	Et (5) + MMA (20)	(i)	2.5 (4h)	96
		(ii)	8.0	213
				(total 309)

without further induction, and simultaneous use of methylammonium and ethanol occurred with a mixture of substrates (Table 4). If, however, strain GF1 was previously sequentially subcultured on ethanol (20 mM), the ability to use methylammonium was repressed and biphasic growth occurred in mixtures of the two substrates (Table 4).

In the above tests, the yields due to each of ethanol and methylammonium were 21.8-25.1 (average 23.5) and 9.8-10.6 (average 10.2), respectively. For MMA-grown inocula the biomass produced was the sum of the separate yields but the growth rate was stimulated essentially to that typical of ethanol alone (Table 4). Using ethanol inocula, the biphasic growth appeared to be due to sequential use of ethanol then MMA, as the yields for the two phases were: "ethanol phase", 20.0 g/mol and "methylammonium phase", 10.5 g/mol, compared to the yields of 21.8 and 10.6 g/mol for the separate substrates.

In these tests, the yields due to each of ethanol and methylammonium were 21.8-25.1 (average 23.5) and 9.8-10.6 (average 10.2), respectively. For MMA-grown inocula the biomass produced was the sum of the separate yields but the growth rate was stimulated essentially to that typical of ethanol alone (Table 4). Using ethanol inocula, the biphasic growth appeared to be due to sequential use of ethanol then MMA, as the yields for the two phases were: "ethanol phase", 20.0 g/mol and "methylammonium phase", 10.5 g/mol, compared to the yields of 21.8 and 10.6 g/mol for the separate substrates.

Concluding remarks

Using *Paracoccus versutus* as a model versatile aerobe, it is clear that esters and ketones as well as low molecular weight alcohols, acids and amines can be degraded aerobically with high efficiency. Cometabolism of alcohols, acids and amines can occur but the prior growth substrates strongly influence the ability of the bacteria to attack some substrates without delay. Use of ethyl formate, methyl acetate, ethyl acetate and methylammonium requires adaptation to those substrates, but adaptation to ethyl acetate also confers ability to use methyl acetate. Methyl formate is a substrate for *Paracoccus versutus*, by virtue of its chemical hydrolysis to its constituent methanol and formate, both of which can be used by some strains of *P. versutus*.

While the esters of short-chain alcohols are relatively minor trace gases in the cycling of carbon, they are certainly compounds with a potentially rapid biological turnover, particularly through degradation by aerobes such as *P. versutus*.

Some of the growth phenomena reported here have significance to the general principles of the 'diauxic' and 'mixotrophic' growth of bacteria. Diauxie is commonly observed when bacteria are challenged with relatively high concentrations of two substrates, for which they need to use different key enzymes in order to grow. Commonly, growth on one substrate is completed before derepression/synthesis of the key enzyme(s) for the second substrate takes place to allow its use for growth. Diauxie, and biphasic growth where there may be some 'overlap' in the consumption of dual substrates, is seen in *P. versutus* growing on a range of mixed substrates, including sugars, organic acids, thiosulfate and methylammonium (Wood and Kelly 1977, 1980, 1983; Kraczkiewicz-Dowjat and Kelly 1985, 1991). Mixotrophy defines the growth mechanisms operating to allow simultaneous use of more than one substrate. In *P. versutus* this can be use of mixtures of heterotrophic substrates or of substrates which separately support either heterotrophic or autotrophic growth (e.g. glucose + formate; Smith et al. 1980; Wood and Kelly 1981). Commonly, simultaneous use of two substrates is seen in chemostat culture in which both are provided in growth-rate-limiting concentrations, in contrast to the diauxie or mixotrophy of batch cultures in which both substrates are initially in excess. The results described here with substrates such as ethyl acetate and mixtures of methylammonium and ethanol are significant to diauxie, biphasic growth and mixotrophy. In the case of alkyl esters such as ethyl acetate a single compound effectively provides the organism with a metabolic challenge in that as soon as it hydrolyzes the substrate it has created a dual substrate environment which may have regulatory impacts on its metabolism. Thus, methyl formate provides two autotrophic substrates, ethyl formate or methyl acetate the heterotrophic + autotrophic "choice" (as methanol and formate support only autotrophic growth in *P. versutus*), and ethyl acetate a pair of heterotrophic substrates.

The response of *P. versutus* when presented with ethyl acetate is thus a series of metabolic phases progressing to full adaptation to use of the compound: first, adaptation of the culture to hydrolyze the substrate, secondly, biphasic use of the two constituent compounds with preferential use of the ethanol moiety, finally full adaptation to simultaneous use (i.e. a single growth rate with no biphasic growth or diauxie) of both ethanol and acetate components. This final state seemed only to be achieved after several serial transfers (and hence a large number of generations) and could represent progressive modification of the phenotypic expression of the genetic system of the whole population, or the selection of variants from within the population with differing regulatory ability. The initial development of cultures able to grow rapidly on methyl acetate or ethyl acetate (presumably by expression of an esterase) was a slow phenomenon (around 140 h for expression of rapid growth on ethyl acetate) and could reflect delayed adaptation by the bulk population or could have resulted from selection of population variants able to produce high levels of specific esterases. This is worthy of further study.

Whether or not mixotrophy or biphasic growth occurred with mixed substrates such as methylammonium and ethanol was here shown to be a function of the previous growth history of the bacteria: *P. versutus* previously grown on ethanol showed diauxie on the mixture, but prior growth on methylammonium enabled mixotrophy (at the "ethanol growth rate", Table 4) when transferred into the dual substrate medium.

In the natural environment the mixotrophic use of such combinations of two or more substrates (and the component moieties of esters) would be much more probable than biphasic or diauxic use, given the low concentrations of any given compound likely to be encountered in situ, especially when esters, ketones and amines are considered as the "trace substrates" of interest, as in the present context of carbon cycle trace gases.

Acknowledgements

We acknowledge support from the (former) Science and Engineering Research Council, which enabled the original work described here to be carried out. We are grateful to P. Whyte who carried out some of the growth experiments with wild type *P. versutus*.

References

Albone ES (1984) Mammalian semiochemistry. The investigation of chemical signals between mammals. Wiley Chichester

Albone ES, Perry GC (1976) Anal sac secretion of the red fox, *Vulpes vulpes*: volatile fatty acids and diamines: implications for a fermentation hypothesis of chemical recognition. J Chem Ecol 2:101-111

Albone ES, Robins SP, Patel D (1976) 5-Aminovaleric acid, a major free amino acid component of the anal sac secretion of the red fox, *Vulpes vulpes*. Comp Biochem Physiol 55B:483-486

Armstrong DW, Martin SM, Yamazaki H (1984) Production of ethyl acetate from dilute ethanol solutions by *Candida utilis* Biotechnol Bioeng 26: 1038-1041

Conkle JP, Camp BJ, Welch BE (1975) Trace composition of human respiratory gas. Arch Environ Health 30:290-295

Darling FMC (1990) Aspects of the semiochemistry of mammalian carnivores. PhD Thesis, CNAA London

Grosjean D (1992) *In situ* organic aerosol formation during a smog episode-estimated production and chemical functionality. Atmos Environ A 26:953-963

Kallelmiri H, Miclo A (1993) Mechanism of ethyl acetate synthesis by *Kluyveromyces fragilis*. FEMS Microbiol Lett 111:207-212

Katayama Y, Hiraishi A, Kuraishi H (1995) *Paracoccus thiocyanatus* sp. nov., a new species of thiocyanate-utilizing facultative chemolithotroph, and transfer of *Thiobacillus versutus* to the genus *Paracoccus* as *Paracoccus versutus* comb. nov. with emendation of the genus. Microbiology (UK) 141:1469-1477

Kelly DP, Wood AP (1982) Autotrophic growth of *Thiobacillus* A2 on methanol. FEMS Microbiol Lett 15:229-233

Kelly DP, Wood AP (1984) Potential for methylotrophic autotrophy in *Thiobacillus versutus* (*Thiobacillus* sp.strain A2). In Crawford RL, Hanson RS (eds) Microbial Growth on C_1 Compounds pp.324-329. American Society for Microbiology, Washington, DC, USA.

Kelly DP, Harrison AP (1989) Genus *Thiobacillus* Beijerinck. In Staley JT, Bryant MP, Pfennig N,and Holt JG (eds) Bergey's Manual of Determinative Bacteriology, pp 1842-1858. Baltimore: Williams and Wilkins.

Kelly DP, Wood AP, Gottschal JC, Kuenen JG (1979) Autotrophic metabolism of formate by *Thiobacillus* strain A2. J Gen Microbiol 114:1-13

Kraczkiewicz-Dowjat A, Kelly DP (1985) Isolation and partial characterization of mutants of *Thiobacillus versutus* deficient in autotrophic metabolism. Microbios 44:185-199
Kraczkiewicz-Dowjat A, Kelly DP (1991) Methylammonium as a substrate for autotrophic growth by wild type *Thiobacillus versutus*, and mutant strains with lesions in chemolithotrophic metabolism. Acta Microbiol Polon 40:159-169
Longo E, Velasquez JB, Sieiro C, Calo P, Villa TG (1992) Production of higher alcohols, ethyl acetate, acetaldehyde and other compounds by 14 *Saccharomyces cerevisiae* wine strains isolated from the same region. World J Microbiol Technol 8:539-641
Lukins S, Foster JW (1963) Methyl ketone metabolism in hydrocarbon-utilising mycobacteria. J Bacteriol 85:1074-1087
Madigan MT (1990) Photocatabolism of acetone by nonsulfur purple bacteria. FEMS Microbiol Lett 71:281-286
Manolis A (1983) The diagnostic potential of breath analysis. Clin Chem 29:5-515
Mattheis JP, Fellman JK, Chen PM, Patterson ME (1991) Changes in headspace volatiles during physiological development of Bisbee Delicious apple fruit. J Agric Fd Chem 39:1902-1906
Meyers AJ (1982) Obligate methylotrophy: evaluation of methyl formate as a C_1 compound. Can J Microbiol 28:1401-1404
Mori,K, Puapoomchareon P (1990) Pheromone synthesis. 118. Synthesis of the enantiomers of (E)-1-ethyl-5-methyl-4-heptenyl acetate (quadrilure), the aggregation pheromone of *Catharus quadricollis*. Liebigs Ann Chem 2:159-162
Morikami T, Tanaka S, Hashimoto Y (1993) Formic and acetic acids in the marine atmosphere. Nippon Kagaku Kaishi 1:98-104
Nakagawa A, Tsujita T, Okuda H (1984a) Purification and some properties of intracellular esterase from *Pseudomonas fluorescens*. J Biochem 95:1047-1054
Nakagawa A, Tsujita T, Okuda H (1984b) Effect of BRIJ-58 on the hydrolysis of methyl butyrate by lipase from *Pseudomonas fluorescens*. J Biochem 96:815-820
Natynczuk S, Bradshaw JWS, MacDonald DW (1989) Chemical constituents of the anal sacs of domestic dogs. Biochem Syst Ecol 17:83-87
Platen H, Schink B (1989) Anaerobic degradation of acetone and higher ketones via carboxylation by newly isolated denitrifying bacteria. J Gen Microbiol 135:883-891
Platen H, Temmes A, Schink B (1990) Anaerobic degradation of acetone by *Desulfococcus biacutus* spec. nov. Arch Microbiol 154:355-361
Price B, Rittenberg SC (1950) The metabolism of acetone. J Biol Chem 185:449-460
Schultztokos JJ, Tanaka S, Morikami T, Shigetani H, Hashimoto Y (1992) Gaseous formic and acetic acids in the atmosphere of Yokohama, Japan. J Atmos Chem 14:85-94
Smith AL, Kelly DP, Wood AP (1980) Metabolism of *Thiobacillus* A2 grown under autotrophic, mixotrophic and heterotrophic conditions in chemostat culture. J Gen Microbiol 121:127-138
Stalder V, Bachofen R (1994) Physiology of growth and carboxyl esterase secretion of a newly-isolated *Bacillus thermoglucosidasius* strain EAEC, enriched on ethyl acetate. Microbiol Res 149:9-16
Taylor DG, Trudgill PW, Cripps RE, Harris PR (1980) The microbial metabolism of acetone. J Gen Microbiol 118:159-170

Whitfield FB, Shea SR, Gillen KJ, Shaw KJ (1981) Volatile components from the roots of *Acacia pulchella* R.Br. and their effect on *Phytophthora cinnamomi* Rands. Aust J Bot 29:195-208
Wood AP, Kelly DP (1977) Heterotrophic growth of *Thiobacillus* A2 on sugars and organic acids. Arch Microbiol 133:257-264
Wood AP, Kelly DP (1978) Triple catabolic pathways for glucose in a fast-growing strain of *Thiobacillus* A2. Arch Microbiol 117:309-310
Wood AP, Kelly DP (1980) Regulation of glucose catabolism in *Thiobacillus* A2 grown in the chemostst under dual limitation by succinate and glucose. Arch Microbiol 128:91-97
Wood AP, Kelly DP (1981) Mixotrophic growth of *Thiobacillus* A2 in chemostat culture on formate and glucose. J Gen Microbiol 125:55-62
Wood AP, Kelly DP (1983) Autotrophic, mixotrophic and heterotrophic growth with denitrification by *Thiobacillus* A2 under anaerobic conditions. FEMS Microbiol Lett 16:363-370
Wood AP, Kelly DP (1983) Use of carboxylic acids by *Thiobacillus* A2. Microbios 38:15-25

Environmental Molecular Biology Approaches: Promises and Pitfalls

Mary E. Lidstrom
Keck Laboratories 138-78
California Institute of Technology
Pasadena, CA 91125

One of the most difficult problems in studying the microbiology of atmospheric trace gas cycling is the inherent uncertainty concerning the makeup of natural populations of these microorganisms. Since many of the relevant microorganisms are difficult to culture, traditional techniques of enumeration and identification that involve culturing are generally unreliable. In addition, since it is clear that a large proportion of the bacteria in nature have never been cultured (Giovannoni et al., 1990; Ward et al., 1990; Liesack and Stackebrandt, 1992), it is not possible to be certain that the bacteria that do grow under laboratory conditions represent the dominant functional population in nature. Determining the functional and phylogenetic diversity of trace gas utilizing or producing populations is an essential step for understanding their role in the dynamics of trace gas flux. This information is also essential for ascertaining how these populations and their relevant trace gas functions (utilization or production) respond to environmental changes. Ultimately, one of the goals of this work is to be able to predict with some certainty how trace gas cycling will be affected by climate change.

Environmental Molecular Biology Approaches

There are several non-culture methods for assessing natural microbial populations, including the use of fluorescent

NATO ASI Series, Vol. I 39
Microbiology of Atmospheric Trace Gases
Edited by J. Colin Murrell and Donovan P. Kelly

antibodies, analysis of signature lipids, and gene probing techniques. Although the first two methods have their advantages for certain applications, the high information content of gene sequences makes the genetic approach attractive as a general technique. Using genetic approaches, a series of levels of information can be obtained. Simple gene probing can provide qualitative information about the presence and relative abundance of specific functional groups. More sophisticated analysis can theoretically provide quantitative information concerning the levels of a relevant population. Finally, if gene sequence data are obtained, comparisons to known sequences can provide information concerning the similarity of key functional genes to those found in well-studied laboratory strains.

The methods used for environmental molecular biology are in general, modifications of standard methods used for molecular genetic analysis of pure cultures. The major hurdle that had to be overcome to extend these methods to environmental samples was the development of high efficiency methods for extracting DNA from natural samples. However, a variety of methods are now available for extracting bulk DNA from sediments, soils and aquatic samples (Fuhrman et al., 1988; Sayler and Layton, 1990; Leff et al., 1995).

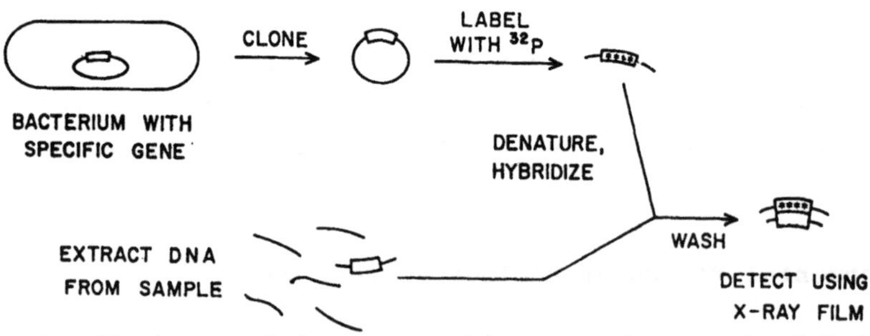

Figure 1. Environmental gene probing. Probe can be labelled with a radioisotope as shown above, or with non-isotopic labelling protocols.

Simple environmental gene probing involves extracting DNA from the sample to be tested, hybridizing the bulk DNA with a labelled probe, and detecting the hybrids, usually with a nitrocellulose or nylon membrane (Fig. 1; Sayler and Layton, 1990). A useful approach is to utilize a slot blot apparatus, in which the DNA is applied to a membrane using a manifold containing a large number of slots of a standard size, and the membrane is then used for hybridization to the probe. Successive dilutions of the sample, compared to positive controls, can provide a relative measure of the amount of DNA detected, as long as the hybridization conditions are controlled (Duong et al. 1994).

This approach is most commonly used with probes in which the target is a gene diagnostic of a particular function, for instance, a gene for the methane monooxygenase to assess methane oxidizing bacteria (Tsien and Hanson, 1992). However, it can also be used with phylogenetic probes specific to the 16S rRNA sequence of a functional group (Sayler and Layton, 1990). In order for this approach to be useful, the probe must be specific, sensitive and highly conserved in the target population. Specificity is best assessed by having sequence data for the target gene from several sources, and comparing conserved sequence areas to sequences in the data base. In addition, positive and negative controls should be carried out to demonstrate that under the conditions used, the probe is both specific and sensitive. It is important to choose a sequence within the target gene that is highly conserved, so that hybridization conditions can be used that discriminate between related genes with a different function and those that carry out the function to be assessed. For instance, the genes for the ammonia monooxygenase and particulate methane monooxygenase are approximately 40% identical at the nucleotide sequence level (Semrau et al, 1995) and probes must be chosen carefully to distinguish between nitrifiers and methanotrophs. One possibility for circumventing this problem is to set the stringency of the hybridization sufficiently high so that only very closely related genes will be detected. However, if the functionally important population contains a more divergent version of the target gene, this approach will not detect that group of bacteria. This line of reasoning points out a fundamental dilemma of this type of gene probing technique, how to discriminate between a small population containing a highly similar gene and a large population containing a divergent gene. Theoretically, both could give the same intensity hybridization signal on a slot blot. One way to address this issue is to carry out a series of hybridizations on each sample with successively increasing stringency. By comparing the loss of signal in the samples to the loss with a suite of positive and negative controls, it should be possible to obtain a qualitative assessment of the gene populations present.

An alternative to direct gene probing is the use of fluorescently-labelled phylogenetic probes to hybridize to rRNA in intact, fixed cells (DeLong et al., 1989). This method allows the direct observation and counting of single cells. In addition, the amount of fluorescence is roughly proportional to the amount of ribosomes in the cell, which is an indication of the physiological state of the cells. The disadvantage of this method is that starved cells found in nature often do not contain sufficient ribosomes to allow fluorescent detection above background. Resuscitation protocols are necessary, that require prior knowledge concerning the physiology of the target group of bacteria. In addition, some groups, for instance the methanogens, exhibit autofluorescence, which can interfere with this procedure. The use of different fluorochromes can theoretically circumvent such problems. The two most commonly used dyes are fluorosceine and rhodamine.

Direct gene probing or direct fluorescent cell detection may not be sufficiently sensitive to detect the target gene if the population containing that gene is low. An alternative method for environmental samples is now being used that involves amplification of the target DNA using the polymerase chain reaction (PCR) (Fig. 2; Sayler and Layton, 1990). This technique requires the presence of small (18-20 base-pair) highly conserved regions within the target gene, which are used to generate primers involved in the amplification reaction. The amplification uses DNA polymerase to make copies of the sequence flanked by the primers. Several rounds of amplification are carried out and the end result is an increase in the amount of the target DNA that is orders of magnitude greater than that present in the initial sample. This amplified DNA can either be probed directly to obtain qualitative information as described above, or it can be cloned and sequenced.

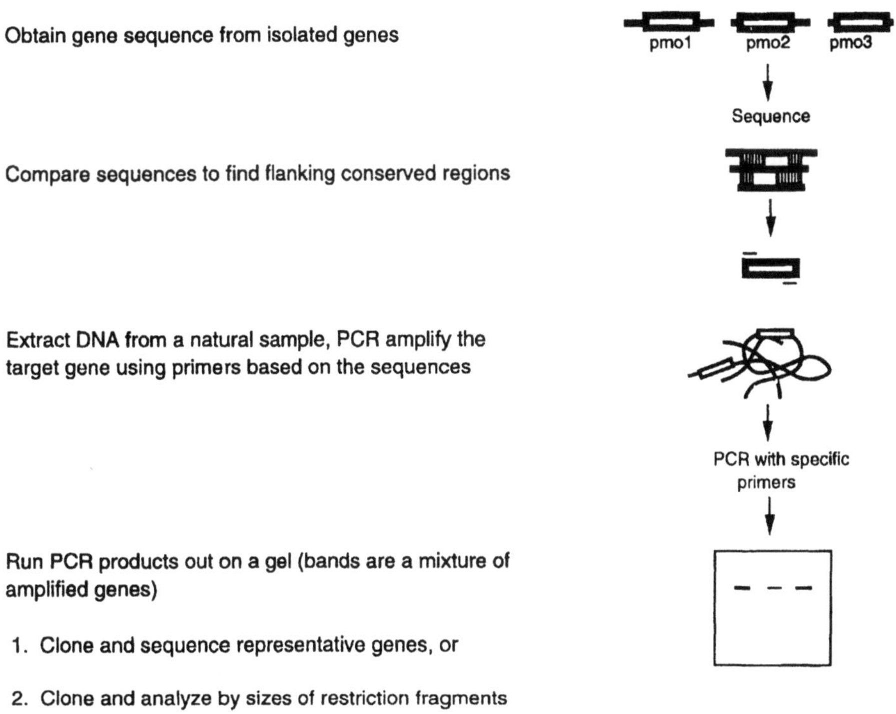

Figure 2. Use of the PCR method to obtain copies of specific DNA fragments from environmental samples for either probing or sequencing.

PCR can be applied to amplification of 16S rRNA genes, since primer sequences are available that will specifically amplify a portion of the genes for known bacterial 16S rRNAs (Britschgi and Giovannoni, 1991; Liesack and Stackebrandt, 1992). The 16S rDNA clones obtained can be screened using probes specific to the sequences of known phylogenetic groups (Britschgi and Giovannoni, 1991; Liesack and Stackebrandt, 1992).

Regardless of whether the PCR-derived clones obtained are from 16S rRNA genes or from diagnostic functional genes, sequence data can be obtained from a number of the clones. These sequences can then be compared and used to generate similarity indices. Not only does this approach increase the database for the target group, it also has the potential to provide information on more divergent genes within the group that may not have been present in strains cultured in the laboratory.

Correlation of Laboratory and Field Data

Most of the approaches noted above depend on the assumption that laboratory cultures are representative of dominant functional microbial groups in the natural environment. The exception to this statement is the approach of using the PCR to amplify 16S rRNA genes, clone them, and sequence a number of them. This approach will only identify bacterial species that are dominant in the ecosystem as a whole, and if the functional group to be identified is a minor part of the bacterial ecosystem in the sample, prescreening with probes is required. Therefore, for most of the bacterial groups involved in utilization or production of trace gases, molecular genetic approaches involve using sequence data or cloned genes from laboratory cultures. As noted in the first paragraph, we cannot be certain that these cultures represent the dominant functional bacterial populations, which places phylogenetic probes in question, nor can we be certain that diagnostic functional genes used for probes are conserved in natural populations. However, an effort can be made to correlate data from field samples with data from laboratory cultures. If a series of key correlations are positive, the chances are good that the dominant functional populations are being identified with probes based on laboratory cultures.

The types of correlations that can be made have to do with the function being measured. Usually, a kinetic analysis can be carried out to determine whether the bulk process being studied has kinetic parameters similar to those found in the laboratory strains. If so, it suggests that the enzyme systems involved may be similar. Such data provide some confidence that the genes encoding those enzyme systems may be conserved. If the kinetics correlate, an independent measure of the population of a functional group can be obtained by measuring V_{max} values of pure cultures grown under a variety of conditions and determining the range for the value of V_{max} per cell. V_{max} values can then be measured for field samples, and an estimated cell population can be calculated. If this estimate correlates with the estimates from culture methods, again, some confidence can be placed in the assumption that the major populations have been identified. Finally, if quantitative data can be obtained using both functional and phylogenetic gene probes, these numbers can be compared to each other and to the calculated values. Such internal checks can either verify the methods, or illustrate that a key population has never been isolated.

A Case Study: Correlation of Laboratory and Field Data for Environmental Assessment of Methanotrophs

One of the groups of bacteria important in trace gas cycling is the methanotrophs, the bacteria that utilize methane under aerobic conditions. These bacteria are traditionally difficult to culture from natural samples and it is relatively common to obtain strains that will not grow on agar plates (Hanson et al, 1992). Therefore, this is one of the groups for which gene probing methods of direct detection are important. The known strains cluster into two major phylogenetic groups within the α- and γ-proteobacteria, respectively, and 16S rRNA probes are available that target each of these groups (Brusseau et al, 1994). In addition, functional gene probes are being developed.

The key enzyme diagnostic of the methanotrophs is the methane oxidizing enzyme, methane monooxygenase. Two biochemically and genetically distinct forms of this enzyme are known. All known strains contain the particulate form (pMMO) and in addition a few strains also contain a soluble form (sMMO) (Dalton, 1992). Genes for the sMMO have been cloned and sequenced, and gene probes have been generated based on sequence comparisons (Murrell, 1994; McDonald et al. 1995). However, the pMMO is the diagnostic enzyme for all methanotrophs, and therefore, it is desirable to develop a functional gene probe based on the pMMO.

The pMMO is unstable and has not been reproducibly purified (Dalton, 1992). However, three membrane polypeptides of 45, 27 and 18 kDa are routinely associated with the pMMO. N-terminal amino acid sequence has been obtained from the 45 kDa pMMO-associated polypeptide from three methanotrophs and used to develop an oligonucleotide probe. Using this probe, a fragment of *Methylococcus capsulatus* DNA was cloned that encoded part of this gene, and cloning and sequencing of overlapping fragments revealed not only the complete gene for the 45 kDa polypeptide (*pmoB*) but also the gene for the 27 kDa polypeptide (*pmoA*) (Semrau et al, 1995). Sequence comparisons showed that these genes showed high similarity at both the nucleotide and amino acid sequence levels to genes encoding the subunits of the ammonia monooxygenase (AMO; *amoA* and *amoB*, respectively). Probing of chromosomal DNA from methanotrophs with *pmoB* and *amoA* probes showed that two copies of these genes were present (Semrau et al., 1995). A portion of the second copy of *pmoB* has been cloned from *M. capsulatus* Bath, and it shows 99% identity to the first copy at the nucleotide sequence level. However, it appears that in other methanotrophs, the two copies are not as highly conserved, as the probing revealed light and dark bands in those cases (Semrau et al., 1995). Efforts are underway to clone and sequence both copies of these genes from other methanotrophs. These sequences will be compared in order to obtain sufficient sequence data to develop both oligonucleotide probes and PCR primers that will be specific to one or both copies of the pMMO, but will not identify strains containing the AMO.

Since the cloning and sequencing of *pmo* genes appears likely to result in the development of effective pMMO probes, it is important to compare methane oxidation kinetics for laboratory strains to those found in field samples to assess how relevant these strains and the genes from them may be for probing natural populations of methanotrophs. We have carried out a series of studies of whole-cell methane oxidation kinetics in four different methanotrophs expressing the pMMO, and have found that the K_S for methane varies depending on the initial copper concentration in the growth medium, with higher copper correlated with a lower K_S value. EPR analyses of membranes from cells grown under these different conditions suggest that two forms of the pMMO can exist, a copper-sufficient form with a low K_S and a copper-insufficient form with a higher K_S (Semrau and Lidstrom, unpublished). Values in the literature for cells grown with copper added to the growth medium and assumed to be expressing the pMMO vary between 2-26 μM (Joergensen and Degn, 1983; Harisson, 1973) and the value for whole cells of *M. trichosporium* OB3b expressing the sMMO has been reported to be 92 μM (Oldenhuis et al., 1991).

We have also determined K_S values for methane oxidation in a number of field samples from marine and freshwater habitats and most range from 3 to 17 μM (Table 1), values that are similar to those observed for pure cultures in the laboratory expressing the copper-sufficient form of the pMMO.
In one case, the Saanich Inlet water column samples, a much lower K_S value was observed (0.2-0.27 μM). These data suggest that two populations of methane-oxidizing organisms exist in natural samples, one with low affinity and one with high affinity for methane. This hypothesis is supported by data reported for other habitats. For instance, in a Danish wetland sediment a K_S value of 2-4 μM was reported (King, 1990) while in a study of a variety of soils, values of 0.03-0.05 μM were reported (Bender and Conrad, 1992). In the latter study, soils that were preincubated with methane before analysis showed biphasic kinetics with two K_S ranges, 0.013-.470 μM and 1.74-

27.9 µM, suggesting the presence of two populations of methane-oxidizing bacteria (Bender and Conrad, 1992).

Table 1. Kinetic parameters for field samples and calculated methanotroph populations based on V_{max} for methane.

Site	Ks (µm)	V_{max} (µmol/l/hr)	Calculated Cells/l[1]
Freshhwater			
Lake Washington sediments[2]	5.2-10.7	26-45	$2.6\text{-}4.5\times10^{10}$
Lake Superior sediments[3]	3.4-5.9	0.42-1.0	$0.4\text{-}1\times10^{9}$
Marine			
Newport Bay estuary sediments[4]	12-17	0.001-0.012	$1\text{-}12\times10^{6}$
Princess Louisa Inlet sediments	6-8	0.004-0.008	$4\text{-}8\times10^{6}$
Saanich Inlet water column[5]	0.2-0.27	0.00015	1.5×10^{5}

[1] Average value 1.0×10^{-9} µmol/hr/cell

Hanson and Wattenberg, 1991	9.0×10^{-9}
Semrau and Lidstrom, unpublished	$0.2\text{-}1.1\times10^{-9}$
Roslev and King, 1995	$1.2\text{-}1.6\times10^{-9}$

[2] Lidstrom and Somers, 1984; Kuivila et al., 1988

[3] Remsen et al., 1989

[4] unpublished data

[5] Ward et al., 1989

Since the higher K_s values are in the same range as those measured for laboratory cultures containing the copper-sufficient pMMO, these populations may be expressing the pMMO and are probably not copper-limited. In most habitats studied it appears that methane oxidation and methanotroph populations are methane-limited (Hanson and Wattenberg, 1991). Therefore, the K_S data suggest that sMMO expression would be detrimental in these methane-limited environments.

The results described above suggest that in environments in which the higher K_S values predominate, it is likely that the gene probes based on laboratory cultures will detect the dominant populations. However, in the environments in which the lower K_S values predominate, it is uncertain whether either the phylogenetic or functional gene probes available will detect the dominant population that oxidizes methane. In these environments it would be very interesting to compare estimates of population size based on gene probing techniques to estimates based on methane oxidation activities.

We have calculated an estimate for the number of methanotrophs per liter of the field sample used, based on an average value of V_{max} per cell obtained from laboratory cultures. This value has been reported to be as high as $9x10^{-9}$ μmol/hr/cell (Hanson and Wattenberg, 1991). Our values for different methanotrophs grown with different copper levels are in the range of $0.2-1.2x10^{-9}$ μmol/hr/cell (Semrau and Lidstrom, unpublished) and values published by Roslev and King (1995) for starved and unstarved cells are in the range of $1.2-1.6x10^{-9}$ μmol/hr/cell. Therefore, we have chosen an average value of $1x10^{-9}$ μmol/hr/cell as an estimate. This estimate should be used with the understanding that the actual range in nature might well be a factor of 5 higher or lower. Using this value, populations of methanotrophs are predicted ranging from $2.6-4.5x10^{10}$ cells per liter for Lake Washington sediments to $1.5x10^{5}$ cells per liter for Saanich Inlet water column samples (Table 1). These numbers generally fall at the high end or within an order of magnitude higher than the methanotroph numbers estimated from similar types of environments using culture techniques (Hanson and Wattenberg, 1991), as expected if these culture methods underestimate the true populations.

These data taken together suggest that the cultures of methanotrophs that have been studied in the laboratory and from which both phylogenetic and functional gene probes have been obtained, may indeed be relevant to certain types of natural environments, especially marine and freshwater sediments.

However, they may not be relevant to methane consumption in the water column of the oceans, nor to certain types of soils. The latter environments are those in which the steady-state levels of dissolved methane in the zones of methane oxidation are very low, usually less than 10 nM. It is important to isolate methanotrophs adapted to low methane concentrations and develop gene probing methods for these strains. All of our attempts to isolate such strains from Saanich Inlet samples using either direct plating under low methane or enrichment by either batch culture or methane-limited chemostats have failed (Lidstrom, unpublished). It is now important to develop quantitative gene probing methods for methanotrophs based on both phylogenetic and functional gene probes to compare populations estimated by the methane oxidation activity approach to those estimated by the genetic probing approach. These comparisons will provide further information about the bacteria that oxidize methane in natural samples, and can then be used to carry out studies of how these populations and their activities change in response to environmental change.

References

Bender M, Conrad R (1992) Kinetics of CH_4 oxidation in oxic soils exposed to ambient air or high CH_4 mixing ratios. FEMS Microbiol Ecol 101:261-270

Britschgi T, Giovannoni SJ (1991) Phylogenetic analysis of a natural marine bacterioplankton population by rRNA gene cloning and sequencing. Appl Environ Microbiol 57:1707-1713

Brusseau GA, Bulygina ES, Hanson RS (1994) Phylogenetic analysis of methylotrophic bacteria based on 16S rRNA sequence. Appl Environ Microbiol 60:626-636

Dalton H (1992) Methane oxidation by methanotrophs, physiological and mechanistic implications. In: Murrell JC, Dalton H (eds) Methane and Methanol Utilizers (pp 85-114) Plenum, New York

DeLong EF, Wickham G, Pace NR (1989) Phylogenetic stains: ribosomal RNA-based probes for the identification of single cells. Science 243:1360-1363

Duong MH, Grant SB, Lidstrom ME (1994) Transfer solution chemistry affects mixed-phase hybridizations. Anal Biochemistry 220:431-433

Fuhrman JA, Comeau DE, Hagstrom A, Chan AM (1988) Extraction from natural planktonic microorganisms of DNA suitable for molecular biological studies. Appl Environ Microbiol 54:1426-1429

Giovannoni SJ, Britschgi T, Moyer CL, Field KG (1990) Genetic diversity in Sargasso sea Bacterioplankton. Nature (London) 345:60-63

Hanson RS, Netrusov AI, Tsuji K (1992) The obligate methanotrophic bacteria *Methylococcus, Methylomonas*, and *Methylosinus*. In: Balows A, Truper HG, Dworkin M, Harder, W, Schleifer K-H (eds) The Prokaryotes Second Edition, (pp 2350-2364) Springer-Verlag, New York

Hanson RS, Wattenberg, EV (1991) Ecology of methylotrophic bacteria. In: Goldberg I, Rokem JS (eds) Biology of Methylotrophs (pp325-348) Butterworth-Heinemann, Boston London Oxford Singapore Sydney Toronto Wellington

Harrisson DEF (1973) Studies on the affinity of methanol- and methane-utilizing bacteria for their carbon substrates. J Appl Bacteriol 36:301-308

Joergensen L, Degn H (1983) Methane oxdiation by *Methylosinus trichosporium* measured by membrane-inlet mass spectrometry. FEMS Microbiol Lett 20:331-334

King G M (1990) Dynamics and controls of methane oxidation in a Danish wetland sediment FEMS Microbiol Ecol 74:309-323

Kuivila KM, Murray JW, Devol AH, Lidstrom ME, Reimers CE (1988) Methane cycling in the sediments of Lake Washington. Limn and Oceanog 33:571-581

Leff LG, Dana JR, McArthur, JV, Shimkets LJ (1995) Comparison of methods of DNA extraction from stream sediments. Appl Environ Microbiol 61:1141-1143

Lidstrom ME, Somers L (1984) Seasonal study of methane consumption in Lake Washington. Appl Environ Microbiol 47:1255-1260.

Liesack W, Stackebrandt E (1992) Occurence of novel groups of the domain *Bacteria* as revealed by analysis of genetic material isolated from an Australian terrestrial environment. J Bacteriol 174:5072-5078

McDonald IR, Kenna EM, Murrell JC (1995) Detection of methanotrophic bacteria in environmental samples with the PCR. Appl Environ Microbiol 61:116-121

Murrell JC (1994) Molecular genetics of methane oxidation. Biodegradation 5:145-159

Oldenhuis R, Oedzes, JY, Van der Waarde JJ, Janssen DB (1991) Kinetics of chlorinated hydrocarbon degradation by *Methylosinus trichosporium* OB3b and toxicity of trichloroethylene. Appl Environ Microbiol 57:7-14

Roslev P, King GM (1995) Aerobic and anaerobic starvation metabolism in methanotrophic bacteria. Appl Environ Microbiol 61:1563-1570

Remsen CC, Minnich EC, Stephens RS, Lidstrom ME (1989) Aerobic methane oxidation in Lake Superior sediments. J Great Lakes Res 15:141-146

Sayler GS, Layton AC (1990) Environmental application of nucleic acid hybridization. Ann Rev Microbiol 44:625-648

Semrau JD, Chistoserdov A, Davagnino J, Kenna E, Holmes, AJ, Finch R, Murrell JC, Lidstrom ME (1995) Particulate methane monooxygenase genes in methanotrophs. J Bacteriol 177:3071-3079

Tsien HC, Hanson RS (1992) Soluble methane monooxygenase component B gene probe for identification of methanotrophs that rapidly degrade trichloroethylene. Appl Environ Microbiol 58:953-960

Ward BB, Kilpatrick KA, Wopat AE, Minnich EC, Lidstrom ME (1989) Methane oxidation in Saanich Inlet during summer stratification. Cont Shelf Res 9:65-75

Ward DM, Weller R, Bateson, MM (1990) 16S rRNA sequences reveal numerous uncultured microorganisms in a natural community. Nature (London) 345:63-65

Molecular Ecology of Methanotrophs

J Colin Murrell, Andrew J Holmes, Ian R McDonald and Elizabeth M Kenna
Department of Biological Sciences
University of Warwick
Coventry CV4 7AL
England

Introduction

Methane oxidizing bacteria (methanotrophs) are unique in growing with methane as their sole source of carbon and energy. They do not grow on multi-carbon compounds, but some can also utilize methanol as a growth substrate. These unique organisms appear to be ubiquitous in the natural environment and have been isolated from a wide variety of soils, sediments and freshwater samples (Whittenbury *et al.* 1970; Bowman *et al.* 1993). There are also marine representatives (reviewed in Murrell and Holmes, 1995). They are all strictly aerobic, gram negative bacteria that grow on a minimal medium and methane and can be classified into two groups on the basis of their intracytoplasmic membranes, pathways of formaldehyde assimilation and 16S rRNA sequence. The five genera *Methylomonas, Methylobacter, Methylococcus, Methylocystis* and *Methylosinus* originally proposed by Whittenbury *et al.* (1970) have largely remained unaltered (Bowman *et al.* 1993). Type I methanotrophs *Methylobacter* and *Methylomonas* are related to bacteria in the γ-subdivision of the *Proteobacteria*, contain bundles of intracytoplasmic membranes and utilize the ribulose monophosphate (RuMP) pathway for formaldehyde assimilation into biomass. Type II methanotrophs, *Methylosinus* and *Methylocystis* lie in the α-subdivision of the *Proteobacteria* have peripheral intracytoplasmic membranes and use the serine pathway for formaldehyde assimilation into biomass. *Methylococcus capsulatus* (Bath) is often referred to as Type X since it has characteristics of both groups. The taxonomy of methanotrophs has been extensively reviewed (Green 1992;

NATO ASI Series, Vol. I 39
Microbiology of Atmospheric Trace Gases
Edited by J. Colin Murrell and Donovan P. Kelly

Bowman 1992; Hanson *et al.* 1993). Largely due to the efforts of Hanson, Bowman and colleagues, an extensive database of 16S rRNA sequences of methanotrophs is available which further strengthens the taxonomy of these organisms (Bratini *et al.* 1992; Brusseau *et al.* 1994).

The interest in methanotrophs over the last 25 years has largely been due to their biotechnological potential for the production of single-cell protein, propylene oxide and other biotransformation products (e.g. Leak 1992). Their use in bioremediation is also the subject of interest since they are able to degrade the groundwater pollutant trichloroethylene (TCE) and other halogenated hydrocarbons (Oldenhuis and Janssen 1993). More recently it has been recognised that methanotrophs are a major sink for atmospheric methane and can also remove much of the methane, produced anaerobically by methanogens, before it enters the atmosphere (recently reviewed in Cicerone and Oremland 1988; Reeburgh 1993; King 1992, 1993). Therefore, these organisms can be considered major players in the biological methane cycle. Aerobic methane oxidation activity is generally maximal at the oxic-anoxic interface of freshwater environments, where the methane and oxygen concentrations are optimal for methanotrophs (Rudd and Taylor 1980). Microbes that oxidize methane under anaerobic conditions have yet to be isolated, however, there is some evidence for their existence in certain environments (e.g. see Chapters by Iversen and Nedwell, this volume).

Physiology, biochemistry and molecular biology of methanotrophs

Methanol produced from the oxidation of methane is further oxidized to formate and CO_2 (which is not usually fixed by ribulose bisphosphate carboxylase) to generate reducing power ($NADH_2$) or it may be assimilated into the cell via the RuMP or serine pathways (e.g. Anthony 1982, 1986). The initial methane oxidation step is carried out by the enzyme methane monooxygenase. The most well-characterized are the soluble-cytoplasmic (sMMO) enzymes from *Methylococcus capsulatus* (Bath) (Dalton *et al.* 1993) and *Methylosinus trichosporium*

OB3b (Lipscomb, 1994). During copper-deficient growth both organisms synthesize a similar sMMO which consists of 3 components. Protein A is the hydroxylase and site of methane binding. It consists of 3 subunits of 60, 45 and 20 kDa arranged in an α_2, β_2, γ_2 configuration and contains a hydroxo-bridged dinuclear iron cluster that is the site of catalysis. Protein B (16 kDa) has no cofactors. Protein C (39 kDa) the reductase component, contains FAD and an [Fe_2S_2] cluster. Not all methanotrophs are able to express sMMO under copper-limited growth conditions, however all appear to contain a membrane-associated, particulate methane monooxygenase (pMMO) when excess copper is available for growth (Dalton, 1992). This pMMO is difficult to purify in active form although polypeptides of 45 and 27 kDa are likely subunits. The high correlation between copper concentration in membranes and pMMO activity, together with EPR studies suggests that a trinuclear copper centre may constitute the active site of pMMO (Chan *et al*; 1993, Nguyen *et al*. 1994).

The genetics and molecular biology of methanotrophs has been extensively reviewed recently (Murrell 1992, 1994; Barta and Hanson, 1993) and will not be discussed in detail here. The sMMO gene clusters have been cloned and DNA sequenced from both *Mc. capsulatus* (Bath) and *Ms. trichosporium* OB3b (reviewed in Murrell 1992, 1994). The genes encoding the α, β and γ subunits of Protein A (*mmoX*, *Y* and *Z*), Protein B (*mmo*B) and Protein C (*mmo*C) are all clustered on the chromosome of both of these organisms. There is a high degree of homology between these genes from these methanotrophs. Screening of these genes and derived amino acid sequences show that with the exception of (i) two small regions of the Protein A α subunit and the diiron centre of ribonucleotide reductase (Protein A also has a diiron centre at its active site) (ii) the N-terminal 90 amino acids of Protein C, which has significant homology with reductase components of several oxygenase enzymes, these genes and the proteins they encode are unique and may be used as indicators for the presence of methanotrophs in the natural environment (see below). The organization, and regulation of these genes has been discussed in detail (Murrell, 1992, 1993, 1994). More recently the genes encoding the pMMO enzyme from *Mc. capsulatus* (Bath) have been cloned and sequenced (Semrau *et al.* 1995). *pmoA* codes for a 27 kDa polypeptide, which has 47% identity with *amoA,* encoding the 27 kDa acetylene binding subunit of ammonia monooxygenase (AMO) from *Nitrosomonas europaea* (McTavish *et al.*, 1993). *pmo*B codes for a 45 kDa polypeptide that has approximately 43% identity with the 43 kDa membrane-bound AmoB of AMO.

Molecular ecology

There have been a number of recent reviews on the cycling of methane in the natural environment and the role that methanotrophs may play in this cycle (e.g. see Rudd and Taylor, 1980; Cicerone and Oremland, 1988; Hanson and Wattenburg, 1991; Kiene, 1991; King, 1992; King, 1993; Reeburgh, 1993; Hanson *et al.* 1993) and therefore this review will be limited to the use of molecular ecological techniques to detect and identify methanotrophs in natural environmental samples, including soils, sediments, freshwater and the marine environment.

The shortcomings of culture-based techniques for microbial ecology are well-documented (e.g. see Ward *et al.* 1992). Even though the classical enrichment and isolation methods of Whittenbury (Whittenbury *et al.* 1970) are excellent for obtaining methanotrophs from environmental samples, many methanotrophs do not grow well on conventional media (e.g. Lees *et al.* 1991) and cultures finally obtained by traditional enrichment techniques may not be the major players in methane cycling in the environment from which they were isolated. Estimates of biomass and activity of methanotrophs in natural samples can be obtained by measuring the evolution of $^{14}CO_2$ and incorporation of $^{14}CH_4$ added to soil or water samples but this will not allow an analysis of the types of bacteria present (e.g. some methane oxidation could be due to ammonia oxidizing bacteria). Another method for identification of methanotrophs in natural environmental samples is phospholipid and fatty acid analysis. This is a relatively expensive technique, however, and would require an extensive data base for methylotrophic organisms. The use of antibody techniques to detect methanotrophs is also potentially an attractive idea and some success has been obtained using antibody raised against whole cells of methanotrophs. Antibodies raised against methanol dehydrogenase from marine methanotrophs and methylotrophs have also been used (reviewed in Murrell and Holmes, 1995). Methods for the use of antibodies raised against key enzymes of methane oxidation such as methanol dehydrogenase and sMMO have also been discussed previously (Hanson and Wattenburg, 1991). However, there are potentially a number of problems in the use of these techniques, including non-specificity of antibody and lack of cell permeability to antibodies.

Application of molecular ecology techniques to methanotrophs has been aided by the sequencing of 16S rRNA genes from a large number of methanotrophs and methylotrophs (Tsuji *et al.* 1990; Bratina *et al.* 1992; Bowman 1992; Murrell *et al.* unpublished). The tight phylogenetic clustering of *Methylomonas*, *Methylobacter* and *Methylococcus* in the γ-*Proteobacteria,* and *Methylosinus* and *Methylocystis* in the α-2 subdivision of the class *Proteobacteria* (Hanson *et al.* 1993; Boulygina *et al.* 1993) suggests that this 16S rRNA sequence analysis may be a good indicator of the unique physiology of this group of organisms. Hanson and colleagues have designed oligonucleotide probes of broad specificity for both serine pathway and RuMP pathway methylotrophs (Tsien *et al.* 1990; Hanson and Wattenberg, 1991; Hanson *et al.* 1993). Five oligonucleotide probes based on analysis of 36 methylotrophic and 10 reference organism 16S rRNA gene sequences have been designed which exploit the observations that methanotrophs are separated into coherent clusters that share common physiological characteristics. When labelled with digoxigenin-11-dd UTP these probes detected different groups of methanotrophs in methane-enriched soil samples. RNA extracted from these soils was shown to contain 16S rRNA seuqences of methanotrophs, when tested by Northern slot blot and hybridization analysis (Brusseau *et al.* 1994).

Restriction Fragment Length Polymorphism (RFLP) in conjunction with hybridization to *mxaF* sequences coding the MDH large subunit structural gene, and part of the sMMO gene cluster, have been used to detect methanotrophs (Tsuji *et al.*

1989; Tsien and Hanson, 1992). sMMO gene probes have also been used to detect *mmo* genes in methanotrophs (Koh *et al.* 1993) including those from TCE-degrading bioreactors (Alvarez-Cohen *et al.* 1992).

As the database of sMMO gene and protein sequences continues to grow, it is clear that several of these sMMO polypeptides appear to be unique when compared with other sequences in current protein data bases. Direct comparison of the complete sMMO gene cluster sequences from *Mc.capsulatus* (Bath) and *Ms.trichosporium* OB3b has allowed the design of oligonucleotide primers that can be used to amplify internal fragments of all five sMMO structural genes from a number of representative methanotrophs in our culture collection. This has added to the database of available sequence information and strengthened the hypothesis that certain sMMO polypeptides are unique (particularly *mmoX*) and therefore may be used as functional gene probes when analysing uncultivated methanotrophs directly from environmental DNA samples. When total bacterial DNA was isolated from a variety of freshwater, soil and peat (wetland) samples and these DNA samples were used as template with the sMMO-specific primers, a PCR product of the correct (predicted) size was obtained. When probed with the appropriate sMMO gene from either *Methylococcus* or *Methylosinus*, a high degree of homology was observed. Interestingly, no sMMO gene sequences have been detected in any marine samples tested to date (Holmes, Murrell, unpublished observations). This is in agreement with all the information in the literature which suggests that only methanotrophs containing a pMMO and not the additional sMMO system are found in the marine environment (Murrell and Holmes 1995; Holmes *et al.* 1995a).

In a more detailed study we have used the sMMO gene/PCR approach to analyse methanotrophs in peat core samples. A 30 cm deep core was extracted from a peatbog at Moorhouse/Upper Teesdale, in the Pennines, UK, and sliced into 1 cm fractions. DNA was subsequently prepared from these samples and used as template DNA in PCR reactions using sMMO-gene primers. All DNA samples yielded PCR products of the correct size with primers for *mmoX* and *mmoC*, suggesting that methanotrophs containing sMMO were distributed through the aerobic and anaerobic (below the level of the water table) zones of the peat bog. Subsequent cloning and DNA sequencing of a number of these PCR products has proved that they are in fact sMMO gene sequences. When compared with the corresponding derived polypeptide sequences from *Methylococcus* and *Methylosinus* it was clear that the *mmoX* gene appears to be the more highly conserved one in these environmental DNA samples with the majority exhibiting approximately 40 to 90% similarity at the amino acid level with known sMMO Protein A α subunit sequences. *mmoC* gene product sequences obtained from peat samples appeared to be more divergent, however, one clone exhibited 100% identity with the corresponding region of the *mmoC* gene from *Ms. trichosporium* OB3b, suggesting that this organism was present in that sample (McDonald *et al.* 1995). These results show that there is considerable diversity of methanotrophs in the peat environment. Using DNA samples extracted from the same peat core and analysing total 16S rRNA gene libraries (see below) with methanotroph group - specific probes it appears that the predominant methanotrophs in this environment may well be *Methylococcus*- and *Methylosinus*-like organisms (McDonald, Murrell, unpublished observations). Enrichment and isolation experiments also lend support to this observation.

The main drawback to the use of sMMO genes as functional probes for methanotrophs is that not all methanotrophs appear to contain the sMMO enzyme system (Stainthorpe *et al.* 1990) but rely solely on the pMMO for growth. The recent cloning and sequencing of the genes encoding the putative 27 and 45 kDa polypeptides of pMMO from *Methylococcus capsulatus* (Bath) (Semrau *et al.* 1995), *pmoA* and *pmoB* respectively, has enabled the use of these genes as DNA probes to identify *pmo* genes in a number of other methanotrophs, including the type I methanotroph *Methylobacter albus* BG8 and the Type II methanotrophs *Ms. trichosporium* OB3b and *Methylocystis parvus* OBBP. Hybridization analysis has also shown that there may be two copies of the *pmo* genes in some methanotrophs (Semrau *et al.* 1995; Lidstrom, Murrell *et al.* unpublished). This has also been noted with the ammonia monooxygenase of *Nitrosomonas europaea* (McTavish *et al.* 1993).

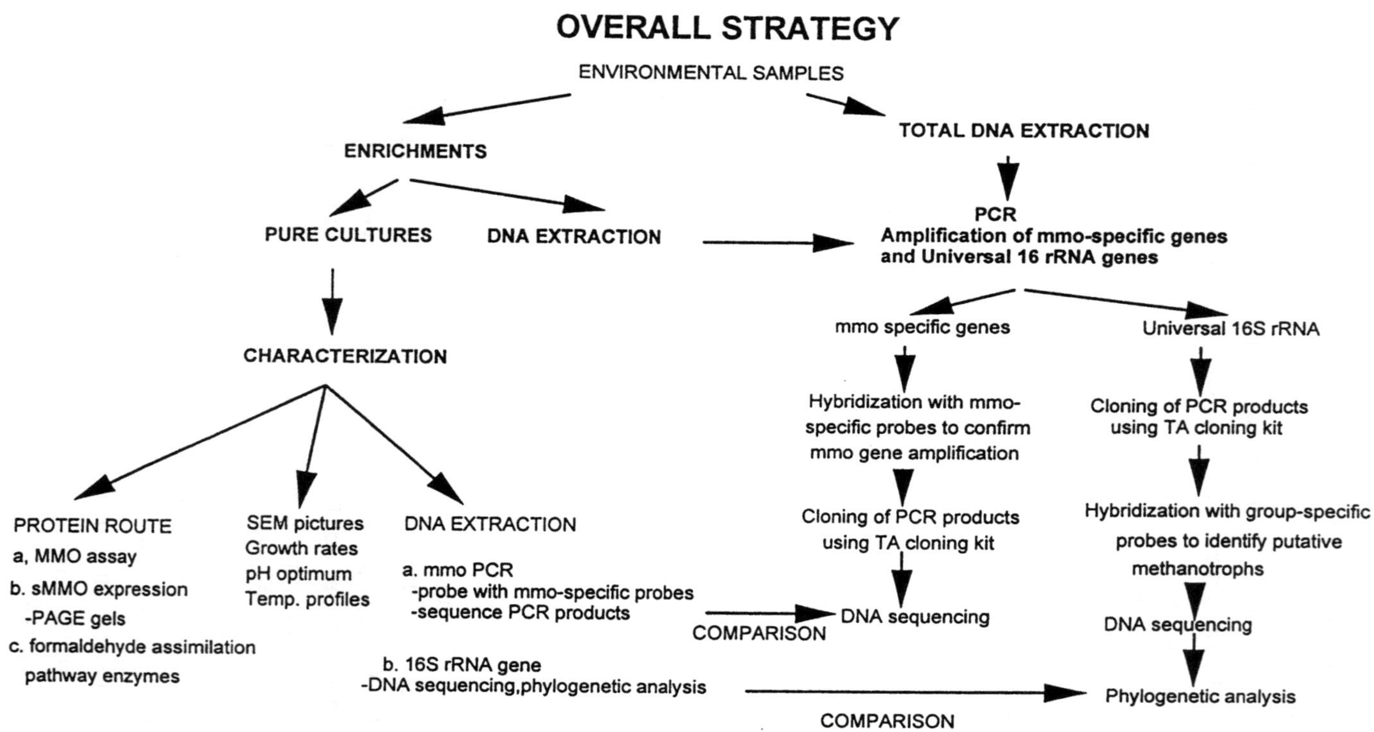

Figure 1. Overall strategy for combining molecular ecology techniques with conventional microbiological techniques for studies on the ecology of methane oxidising bacteria from a variety of different environments

The availability of *pmo* gene sequences has allowed us to design degenerate oligonucleotide primers based on the derived amino acid sequence of the 27 kDa polypeptide of pMMO for the PCR amplification of *pmoA*. A 525 bp internal DNA fragment of *pmoA* from several methanotrophs including *Mc. capsulatus* (Bath), *Ms. trichosporium*, *Methylomonas methanica*, *Methylocystis parvus* and *Methylobacter albus* has now been amplified by PCR, cloned and sequenced. In addition, the same degenerate primers can be used to amplify the corresponding *amoA* gene sequences from a variety of nitrifiers. (No PCR products were obtained with DNA from negative control bacteria that do not oxidize methane or ammonia). In this comparative study, representatives of each of the phylogenetic groups of methanotrophs (α- and γ- *Proteobacteria*) and ammonia oxidizing nitrifying bacteria (β- and γ-*Proteobacteria*) were included (**Figure 2**). Analysis of the predicted amino acid sequences from these gene fragments revealed strong conservation in primary structure (Fig. 2). Overall amino acid sequence conservation between AmoA and PmoA was approximately 40% identity (65% similarity). The most notable feature of sequence comparisons was that the level of conservation showed a stronger correlation with the phylogenetic relatedness of the organisms than with their physiology. Three identity groups were observed, representing the α, β and γ sub-divisions of the *Proteobacteria* rather than two identity groups representing AmoA and PmoA (Holmes *et al.* 1995b) (Fig. 3). Significantly, *Nitrosococcus oceanus* AmoA showed greater identity (>75%) with all γ-*Proteobacteria* methanotroph (Type I) PmoA sequences than with any of the β-*proteobacteria* nitrifier AmoA sequences (> 50%), suggesting that ammonia and methane monooxygenases share a common ancestry. From these and other *pmo* gene sequences currently being analysed in the authors' and Prof Mary Lidstrom's laboratories it should now be possible to redesign PCR primers (from regions of amino acid sequence that are more highly conserved in methanotroph PmoA and PmoB) that will selectively amplify only *pmo* gene sequences from the environment.

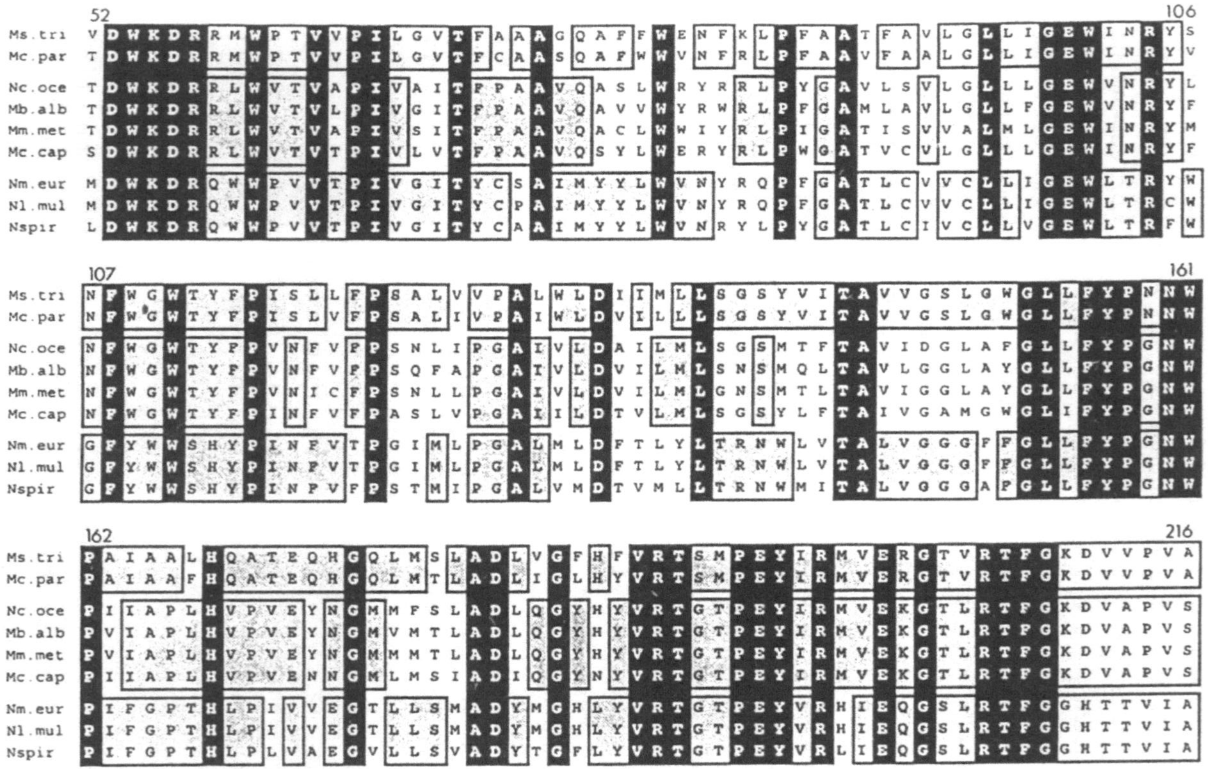

Figure 2. Alignment of predicted amino acid sequences of *pmoA* and *amoA* genes from: Ms. tri, *Methylosinus trichosporium*; Mc. par, *Methylocystis parvus*; Nc. oce, *Nitrosococcus oceanus*; Mb. alb, *Methylobacter albus*; Mm. met, *Methylomonas methanica*; Mc. cap, *Methylococcus* capsulatus; Nm. eur, *Nitrosomonas europaea*; Nl. mul, *Nitrosolobus multiformis*; Nspira, *Nitrosospira* sp. Residues boxed in grey are conserved throughout all members of a subdivision of the *Proteobacteria*. Residues boxed in black are universally conserved in the sequences included in this study. Amino acids are numbered according to previously published sequences for *Mc. capsulatus* (Semrau *et al.* 1995).

	Ms.tri	Mcy.par	Nc.oce	Mb.alb	Mm.met	Mc.caps	Nm.eur	Nl.mul	Nspira
Ms. trichosporium		88%	62%	61%	60%	64%	44%	44%	46%
Mcy. parvus	95%		65%	65%	63%	64%	46%	46%	47%
Nc. oceanus	82%	82%		82%	85%	78%	51%	51%	54%
Mb. albus	80%	81%	92%		82%	75%	49%	49%	49%
Mm. methanica	82%	82%	95%	93%		75%	50%	49%	49%
Mc. capsulatus	83%	80%	92%	89%	91%		54%	54%	53%
Nm. europaea	68%	67%	70%	68%	71%	73%		99%	83%
Nl. multiformis	69%	68%	71%	68%	71%	74%	99%		83%
Nitrosospira sp.	70%	70%	76%	72%	74%	78%	91%	91%	

Figure 3. Identity/similarity matrix derived from the comparison of 169 amino acids of methanotroph PmoA and nitrifier AmoA sequences. Values in the upper right triangle of the matrix are percentage identity values. Those in the lower left triangle are percentage similarity values. Identity/similarity groups consisting of: (A) alpha-*Proteobacteria* methanotrophs (*Methylosinus trichosporium* and *Methylocystis parvus*; (B) gamma-*Proteobacteria* methanotrophs (*Methylobacter albus, Methylomonas methanica* and *Methylococcus capsulatus*) and nitrifiers (*Nitrosococcus oceanus*); and (C) beta-*Proteobacteria* nitrifiers (*Nitrosomonas europaea, Nitrosolobus multiformis* and *Nitrosospira sp.*) are boxed.

16S rRNA genes as phylogenetic probes for methanotrophs

The availability of 16S rRNA gene sequences for methanotrophs has allowed the construction of several sets of oligonucleotide probes designed to identify different groups of methanotrophs and methylotrophs in environmental samples (e.g. Brusseau *et al.* 1994). We have utilized such sequence information to design 16S rRNA probes targeting different genera of methanotrophs (Holmes *et al.* 1995a). Currently we are using these 16S rRNA probes to screen, by colony hybridizations, 16S rRNA gene libraries that have been constructed from DNA extracted from two agricultural soil samples using PCR and the universal eubacterial 16S rRNA primers (Giovannoni, 1991). These libraries (8,000 clones in each library) were from Rothamsted Research Station agricultural soils, which were identical except that one had received a substantial amount of inorganic nitrogen on an annual basis for over 150 years. Comparing the two libraries, we hope to analyse the shift in populations of methanotrophs (and nitrifers) in response to ammonium fertilization since ammonium clearly has an effect on the capacity of soils to act as a sink for atmospheric methane (see Chapters by Steudler, King, Nedwell, this volume). In this type of experiment, careful controls, using cloned methanotroph 16S rRNA genes from representative genera on each set of colony blots to establish optimum hybridization conditions for each probe, are essential. Even then, putative positive methanotroph 16S rRNA gene clones that are "lit-up" with each of the probes need to be sequenced and a phylogenetic analysis carried out to confirm their identity (Kenna and Murrell, unpublished). A similar strategy has been adopted with 16S rRNA gene libraries made from DNA isolated at five depths in the peat core sample described earlier. After probing these libraries with our group-specific methanotroph oligonucleotides, the predominant putative methanotroph 16S rDNA clones appear to be *Methylococcus* and *Methylosinus*. Again, a number of these genes are being sequenced and a phylogenetic analysis carried out to verify their authenticity (McDonald, Murrell, unpublished). The general strategies being used to combine molecular ecological techniques with conventional microbiological methods for examining the ecology of methanotrophs are outlined in Fig. 1.

One of the major limitations of rRNA-targeted group-specific probes is that they may cross-react with organisms of other physiological or phylogenetic groups when applied to environmental samples containing unknown sequences. An extensive database and series of positive and negative controls need to be established for this type of work. The restricted physiology of methanotrophs has been exploited to determine the specificity and efficiency of 16S rRNA probes for these organisms. The approach we have taken is to enrich for methanotrophs in seawater samples and monitor the shift in bacterial populations after enrichment by analysing 16S rRNA gene libraries. A 16S rDNA sequence that was abundant only after methane enrichment was recovered, sequenced and found to show a close phylogenetic relationship to *Methylomonas*. Subsequently a fluor-labelled oligonucleotide probe was synthesized which was specific for this *Methylomonas* 16S rDNA and organisms containing this sequence were observed in methane/seawater enrichments by *in situ* hybridization experiments and fluorescence microscopy (Holmes *et al.* 1995a). Furthermore, it was possible to monitor, by fluorescence microscopy, abundance of this novel marine *Methylomonas* sp. in flask cultures containing different nutrient regimes, thus enabling the optimization of growth conditions to aid isolation (Murrell and Holmes, 1995; Holmes *et al.* 1995a).

Conclusions and future areas for research

The cloning of genes encoding the particulate methane monooxygenase enzyme system of methanotrophs will allow us to achieve a rapid expansion of the database of *pmo* gene sequences. However, before functional gene probes or PCR primers specific for these genes can be designed, pmo genes must be compared with a number of amo gene (and gene product) sequences from different nitrifiers since it is likely that both methanotrophs and nitrifiers will be present in similar environments. Our preliminary sequence comparisons (Fig. 2) suggest that it will soon be possible to design PCR primers that specifically amplify *pmo* genes from both cultures and environmental DNA samples. The 16S rRNA technology described above also has great potential for the identification and subsequent isolation of novel methanotrophs from the environment. One area of interest concerns the bacterium or consortium that is carrying out anaerobic methane oxidation (see Chapters by Iversen, Nedwell, this volume) and it may be that molecular ecology techniques, in combination with conventional microbiological techniques may aid the identification of these types of organisms which have not yet (to the authors' knowledge) been isolated in the laboratory.

Acknowledgements

Work on the molecular ecology of methanotrophs in the authors' laboratory is funded by NERC, BBSRC and the EU.

References

Alvarez-Cohen L, McCarty PL, Boulygina E, Brusseau G, Hanson RS (1992) Cometabolic biotransformation of trichloroethylene and chloroform by a bacterial consortium grown with methane Appl Environ Microbiol 58: 1886-1893

Anthony C (1982) The Biochemistry of Methylotrophs. Academic Press, London

Anthony C (1986) Bacterial oxidation of methane and methanol Adv Micro Physiol 27:113-210

Barta TM and Hanson RS (1993) Genetics of methane and methanol oxidation in gram negative methylotrophic bacteria Anton van Leeuw 64:109-120.

Boulygina ES, Chumakov KM, Netrusov AI (1993) Systematics of gram-negative methylotrophic bacteria based on 5S rRNA sequences. In Murrell JC, Kelly DP (eds) Microbial Growth on C_1 Compounds, pp 275-284. Andover: Intercept Ltd.

Bowman JP, Sly I, Nicholas PD, Hayward AC (1993) Revised taxonomy of the methanotrophs: description of Methylobacter gen. nov., emendation of *Methylococcus*, validation of *Methylosinus* species and a proposal that the family *Methylococcaceae* includes only group I methanotrophs. Int J Syst Bacteriol 43:735-753

Bowman JP (1992) The systematics of methane-utilizing bacteria. PhD Thesis University of Queensland, Brisbane, Australia

Bratina BJ, Brusseau GA, Hanson RS (1992) Use of 16S rRNA analysis to investigate phylogeny of methylotrophic bacteria. Int J System Bacteriol 42: 645-648

Brusseau GA, Bulygina ES, Hanson RS (1994) Phylogenetic analysis and development of probes for differentiating methylotrophic bacteria. Appl Environ Microbiol 60 626-636
Chan SI, Nguyen H-HT, Shiemke AK, Lidstrom ME (1993) Biochemical and biophysical studies towards characterization of the membrane-associated methane monooxygenase In: Murrell JC, Kelly DP (Eds) Microbial Growth on C_1 Compounds (pp 93-107) Intercept Press, Andover, UK
Cicerone RJ, Oremland RS (1988) Biogeochemical aspects of atmospheric methane. Global Biogeochem. Cycles 2:299-327
Dalton H (1992) Methane oxidation by methanotrophs, physiological and mechanistic implications. In: Murrell JC, Dalton H (Eds) Methane and Methanol Utilizers pp 85-114 Plenum, New York
Dalton H, Wilkins PC & Jiang Y (1993) Structure and mechanism of action of the hydroxylase of soluble methane monooxygenase. In: Murrell JC, Kelly DP (Eds) Microbial Growth on C_1 Compounds (pp 65-80) Intercept Press, Andover, UK
Giovannoni SJ (1991) The polymerase chain reaction. In Stackebrandt E, Goodfellow M (eds) Nucleic Acid Techniques in Bacterial Systematics, pp 177-203. Chichester: John Wiley and Sons
Green PN (1992) Taxonomy of methylotrophic bacteria. In: Murrell JC, Dalton H (Eds) Methane and Methanol Utilizers (pp 23-84) Plenum, New York
Hanson RS, Wattenberg EV (1991) Ecology of methylotrophic bacteria. In: Goldberg I, Rokem JS (Eds) Biology of methylotrophs pp 325-348. Butterworth-Heinemann, Boston, USA
Hanson RS, Bratina BJ, Brusseau GA (1993) Phylogeny and ecology of methylotrophic bacteria. In: Murrell JC, Kelly DP (Eds) Microbial Growth on C_1 Compounds pp 285-302. Intercept Press, Andover, UK
Holmes AJ, Owens NJP, Murrell JC (1995a) Detection of novel marine methanotrophs using phylogenetic and functional gene probes after methane enrichment. Microbiology 141: 1947-1955
Holmes AJ, Costello A, Lidstrom ME, Murrell JC (1995b) Evidence that particulate methane monooxygenase and ammonia monooxygenase are evolutionarily related enzymes. FEMS Microbiol Lett, in press
Kiene RP (1991) Production and consumption of methane in aquatic systems In: Microbial Production and Consumption of Greenhouse Gases: Methane, Nitrogen Oxides and Halomethanes (JE Rogers, WB Whitman; Eds) pp 111-146 American Society for Microbiology, Washington DC
King GM (1992) Ecological aspects of methane oxidation, a key determinant of global methane dynamics. Adv Microb Ecol 12:431-468
King GM (1993) Ecophysiological characteristics of obligate methanotrophic bacteria and methane oxidation in situ. In: Murrell JC, Kelly DP (Eds) Microbial Growth on C_1 Compounds pp 303-313. Intercept Press, Andover, UK
Koh S-C, Bowman JP, Sayler GS (1993) Soluble methane monooxygenase production and trichloroethylene degradation

by a type 1 methanotroph *Methylomonas methanica* 68-1. Appl Environ Microbiol 59:960-967

Leak DJ (1992) Biotechnological and applied aspects of methane and methanol utilizers. In: Murrell JC, Dalton H (Eds) Methane and Methanol Utilizers pp 245-279. Plenum Press, New York

Lees V, Owens NJP, Murrell JC (1991) Nitrogen metabolism in marine methanotrophs. Arch Microbiol 157:60-65

Lipscomb JD (1994) Biochemistry of the soluble methane monooxygenase. Ann Rev Microbiol 48:371-399

McDonald IR, Kenna EM, Murrell JC (1995) Detection of methanotrophic bacteria in environmental samples with the PCR. Appl Environ Microbiol 61:116-121

McTavish HJ, Fuchs JA, Hooper AB (1993) Sequence of the gene coding for ammonia monooxygenase in *Nitrosomonas europaea*. J Bact 175:2436-2444

Murrell JC (1992) Genetics and molecular biology of methanotrophs. FEMS Microbiol Rev 88:233-248

Murrell JC (1993) Molecular biology of methane oxidation pp. 109-120 In: Microbial Growth on C_1 Compounds (JC Murrell, DP Kelly, Eds.). Intercept, Andover, UK

Murrell JC (1994) Molecular genetics of methane oxidation. Biodegradation 5:145-159

Murrell JC, Holmes AJ (1995) Molecular ecology of marine methanotrophs. In: Molecular Ecology of Aquatic Microbes, (I. Joint, ed.) NATO ASI Series, Springer-Verlag. In Press.

Nguyen HT, Shiemke AK, Jacobs SJ, Hales BJ, Lidstrom ME, Chan SI (1994) The nature of the copper ions in the membranes containing the particulate methane monooxygenase from *Methylococcus capsulatus* (Bath). J Biol Chem 269:14995-15005

Oldenhuis R, Janssen DB (1993) Degradation of trichlorethylene by methanotrophic bacteria. In: Murrell JC, Kelly DP (Eds) Microbial Growth on C_1 Compounds pp 121-133. Intercept Press, Andover, UK

Reeburgh WS (1993) The role of methylotrophy in the global methane budget. In: Murrell JC, Kelly DP (Eds) Microbial Growth on C_1 Compounds pp 1-14. Intercept Press, Andover, UK

Rudd JWM, Taylor CD (1980) Methane cycling in aquatic environments. Adv Aquat Microbiol 2:77-150.

Semrau JD, Chistoserdov A, Lebron J, Costello A, Davagnino J, Kenna E, Holmes AJ, Finch R, Murrell JC, Lidstrom ME (1995). Particulate methane monooxygenase genes in methanotrophs. J Bacteriol 177:3071-3079

Stainthorpe AC, Salmond GPC, Dalton H, Murrell JC (1990) Screening of obligate methanotrophs for soluble methane monooxygenase genes. FEMS Microbiol Lett 72:345-348

Tsien HC, Bratina BJ, Tsuji K, Hanson RS (1990) Use of oligonucleotide signature probes for identification of physiological groups of methylotrophic bacteria. Appl Environ Microbiol 56:2858-2865

Tsien HC, Hanson RS (1992) A soluble methane monooxygenase component B gene probe for the identification of methanotrophs that rapidly oxidize trichloroethylene. Appl Environ Microbiol 58:953-960

Tsuji K, Tsien HC, Bratina B, Bastien C, Zhang Y, Machlin S, Hanson RS (1989) Genetic and biochemical studies of methylotrophic bacteria. IGT Press, Chicago, USA

Tsuji K, Tsien HC, Hanson RS, De Palma SR, Scholtz R, La Roche S (1990) 16S ribosomal RNA sequences analysis for determination of phylogenetic relationship among methylotrophs. J Gen Microbiol 136:1-10

Ward DM, Bateson MM, Weller R, Ruff-Roberts AL (1992) Ribosomal RNA analysis of microorganisms as they occur in nature. Adv Microb Ecol 12:209-286

Whittenbury R, Phillips KC, Wilkinson JF (1970) Enrichment, isolation and some properties of methane-utilizing bacteria. J Gen Microbiol 61:205-218

Transformation of N_2O and CH_4 in Stratified Microbial Communities Studied by Use of Microsensors

Lars Riis Damgaard
Niels Peter Revsbech
Department of Microbial Ecology
Institute of Biological Sciences
University of Aarhus, Bd. 540
DK-8000 Aarhus C, Denmark

Introduction

The processes producing and consuming the trace gases N_2O and CH_4 are often taking place in microenvironments characterized by steep chemical gradients. The chemical microenvironment and the transformation rates within such stratified communities can be studied by use of microsensors, which can resolve chemical gradients with a spatial resolution of a few micrometers. Reviews on the general use of microsensors in microbial ecology have previously been published (Revsbech and Jørgensen 1986, Revsbech 1994). At present, microsensors for both N_2O (Revsbech et al. 1988) and CH_4 (LR Damgaard and NP Revsbech, unpublished) have been developed, but the N_2O microsensor has almost exclusively been used to quantify denitrification while N_2O reductase is inhibited with acetylene, and the CH_4 microsensor is so new that no work using this sensor has been published. In the present paper we describe the limited work on trace gas transformation presently done by use of microsensors, but it should be emphasized that this is only the beginning of a very detailed description of these processes now enabled by microsensor development.

Transformation of N_2O

<u>Combined microsensor for O_2 and N_2O</u>. Microscale resolution of

NATO ASI Series, Vol. I 39
Microbiology of Atmospheric Trace Gases
Edited by J. Colin Murrell and Donovan P. Kelly

N_2O distribution and thus of denitrifying activity was made possible by the development of a microsensor for combined measurement of O_2 and N_2O (Revsbech et al. 1988). The sensor (Fig. 1) consists of a common casing containing 3 cathodes and with a silicone membrane in the tip. Gases diffuse through the membrane, and oxygen is reduced at the foremost cathode. N_2O is not reduced at this cathode, as such a reduction is dependent on a partially oxidized silver surface (Ag^+ ions are necessary for catalytic activity), and N_2O therefore penetrates to the silver cathode where it is reduced. The backmost cathode is a guard, preventing the diffusion of oxygen and N_2O towards the tip from the bulk reservoir of electrolyte. It is made of porous gold which is subsequently silver plated. This particular microsensor (Fig. 1) is, unfortunately, difficult to make, although some advances in the construction procedure have been made (Dalsgaard and Revsbech, 1992) since the first description was published. A major problem with the construction has been that the foremost gold cathode was dissolved when the N_2O cathode was coated with silver. Silver was applied by injecting a cyanide-complexed solution of silver ions into the casing and then electroplating at about -0.5 V relative to a silver anode. We have now realized that the dissolution of gold during this process can be prevented by doing all electroplatings in the absence of oxygen. When there is no dissolution of gold, the foremost cathode consisting of a porous mass of gold filling the entire lumen of the casing can be made shorter, and it is then possible to make more rapidly responding sensors. The lifetime of N_2O sensors is unfortunately limited to a few days of operation because of the need for a highly alkaline electrolyte (pH 13) which causes rapid degradation of thin glass layers and silicone membrane. There are also some unresolved problems with respect to obtaining maximum sensitivity, and it is often necessary to reverse the polarization voltage at intervals of a few hours to obtain a partially oxidized silver cathode with sufficient catalytic

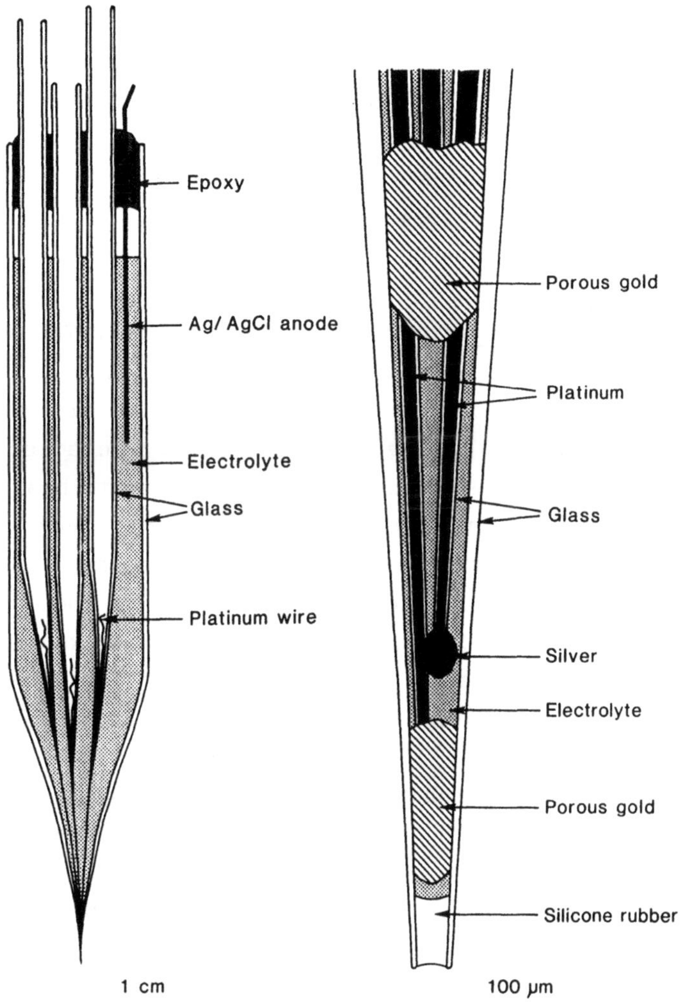

Fig. 1. Microsensor for combined determination of O_2 and N_2O.

activity. For use in denitrification studies, the most disturbing problem has been a sensitivity to acetylene by most sensors, making it difficult to analyze for N_2O in acetylene-inhibited systems. Having mentioned all these problems, it should, however, be stated that the best sensors of this kind work just about ideally and can yield data which could not be obtained by any other means. The sensitivity of such good sensors can be better than 1 µM, and the response is perfectly linear without any interferences except for high sulfide concentrations which cause permanent inactivation of the sensor. The majority of the problems mentioned can, without any doubt, be solved by a careful study, and we actually plan to do such a study in the near future as the development of a microscale NO_3^- biosensor using a N_2O microsensor as sensing element (Damgaard et al. in press, L.H. Larsen and N.P. Revsbech, submitted manuscript) has made it necessary for us to focus on optimal performance.

Transformation of N_2O around hot-spots in soil. The very first microscale determinations of N_2O showed that measurable N_2O profiles could be found in biofilms and similar environments (Revsbech et al. 1988). The only relatively detailed study of N_2O in a non-acetylene inhibited environment has, however, been carried out on organic hot-spots in soil (Højberg et al. 1994). One type of hot-spot was formed by adding a clover leaf to the surface of a soil aggregate containing about 2 mM NO_3^- in the porewater and then allowing the leaf to decompose over a period of about 1 week. During the incubation the aggregate was kept at a water potential near field capacity. Profiles of O_2 and N_2O as a funtion of distance from the leaf center with and without acetylene inhibition of nitrous oxide reductase is shown in Fig. 2. The penetration of oxygen was very superficial in the leaf, whereas oxygen penetrated to a depth of 2-4 mm in the soil, 3-6 mm from the edge of the leaf. Nitrous oxide accumulated while no acetylene was added, although a detectable level was present only below a depth of 5 mm in the profile from the center of the leaf, and below a depth of 2 mm at the edge of the leaf. Maximum levels at depth in the aggregate were about 40 µM. When acetylene was added, the concentrations of N_2O rose dramatically, and a peak concentration was found at 2 mm depth below the center of the leaf where no N_2O was present before acetylene addition. The majority of the N_2O formed in the absence of acetylene was evidently reduced further to N_2. The absence of detectable N_2O below the center of the leaf could be due to local depletion of NO_3^- or to a relatively higher activity of N_2O reductase in this layer as compared to the layers with detectable N_2O.

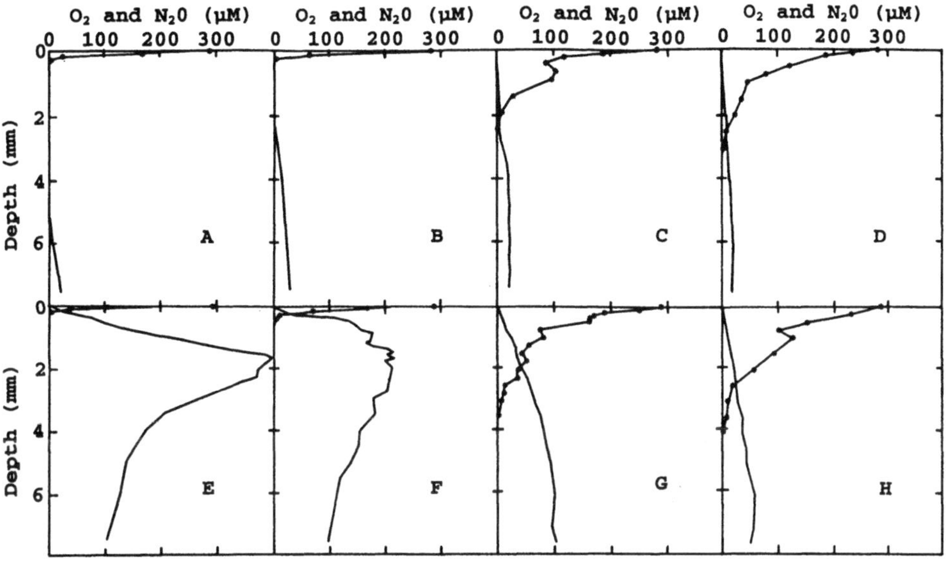

Fig. 2. Profiles of O_2 (curves with symbols) and N_2O (curves without symbols) in a soil aggregate measured at 0 (A and E), 3(B and F), 6 (C and G), and 9 mm (D and H), respectively, from the center of a decaying clover leaf placed on the surface of the aggregate. The leaf was about 6 mm in diameter and 0.3 mm thick. A to D were measured without acetylene addition while E to H were measured after a further 24 h incubation with 1 kPa acetylene.

Addition of excess NO_3^- (not shown) did stimulate N_2O formation, but the appearance of the N_2O peak at only 2 mm depth when acetylene was added indicated that NO_3^- was present at least up to this level without addition of excess NO_3^-.

It is difficult to do detailed calculations of transformation rates based on concentration profiles from soil, as soil, almost by definition, is a very heterogeneous environment where one-dimensional diffusion-reaction models only poorly simulate the actual situation. It is different in water-saturated environments, where modelling may yield even very detailed information about the microdistribution of activities (Christensen et al. 1989, Dalsgaard and Revsbech 1992).

By the study illustrated in Fig. 2 it was difficult to describe the regulating factors for net N_2O formation in detail because we did not have a NO_3^- microsensor which can be used in such an environment. The NO_3^- microsensors described

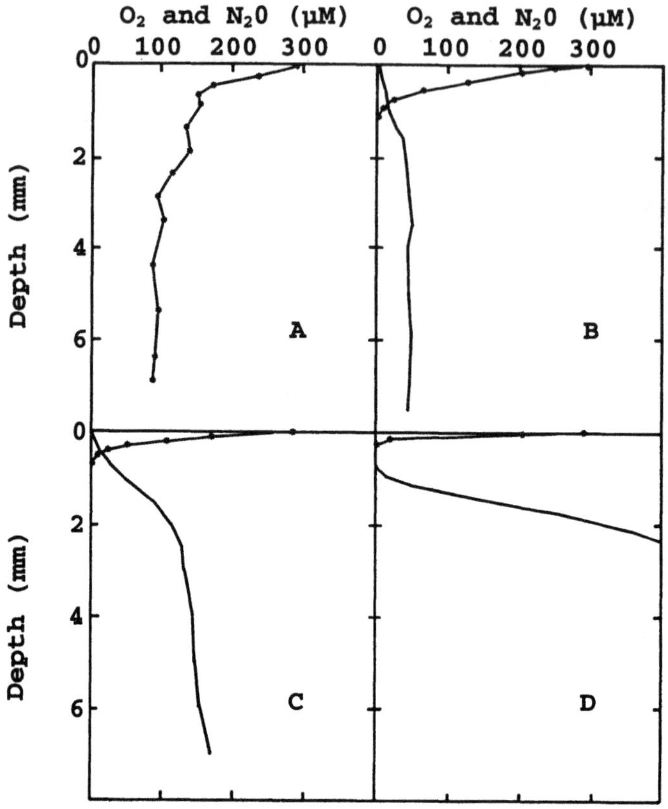

Fig. 3. Profiles of O_2 (curves with symbols) and N_2O (curves without symbols) measured at the same spot of a soil aggregate, but at 0 (A), 2.25 (B), 11 (C) and 19.5 h (D), respectively, after addition of about 25 mg of powderized tryptone to 15 mm^2 of the aggregate surface. No acetylene was added.

until now are based on an NO_3^--ion exchanger embedded in a membrane (de Beer and Sweerts 1989, Jensen et al. 1993), and it has turned out to be impossible to make NO_3^--exchangers with sufficient specificity to avoid interferences from other small anions such as Cl^- and HCO_3^-. The only almost ideal electrode based on such a membrane is the ordinary glass pH electrode, although low levels of interference are also known here. The interference from Cl^- on the previously described NO_3^- microsensors prevents the use of such sensors in marine areas, and the interference from HCO_3^- makes it difficult to analyze any substrate with a high rate of heterotrophic metabolic activity releasing inorganic carbon. The analysis of hotspots with a very intense rate of decompositon is thus not possible using such sensors. In our laboratory we have, however, just succeded in making a NO_3^- biosensor containing immobilized denitrifying bacteria (Damgaard et al. in press, L.H. Larsen and N.P. Revsbech, submitted manuscript) which works without such disturbing interferences and which has a linear response down to less than 1 µM. By use of the new sensor it should thus be possible to study the processes of nitrification and denitrification in all enviroments, including marine areas, and we may then also learn more about the regulating factors for release of N_2O.

The N_2O profiles associated with a decaying clover leaf illustrated in Fig. 2 were recorded after one week of leaf decomposition. It has been described that the time sequence of synthesis of denitrifying enzymes may cause a pronounced net formation of N_2O just after induction of anaerobiosis (Firestone and Tiedje 1979), so a time sequence of N_2O profiles in a soil aggregate was analyzed after the addition of organic matter in the form of 25 mg of tryptone powder (Fig. 3). Before tryptone was added, the aggregate was oxic in all layers, and no N_2O was found. After only 2.25 h of incubation with tryptone, all layers (down to the maximum analyzed depth of 7 mm) below about the 15 mm^2 large hot-spot were anoxic, and N_2O was present at concentrations up to 50 µM. After 19.5 h the concentration of N_2O had increased to about 400 µM at a depth of 2.5 mm, but N_2O was then not present in the uppermost 0.8 mm of the aggegate. Seen as an export to the surrounding atmosphere, there was a net production of N_2O both at 2.25 h and at 11 h, but this production had ceased at 19.5 h. The possible reasons for a layer devoid of N_2O just below a surface-associated hot-spot were discussed above for the clover leaf hot-spot of Fig. 2.

It is clear from the discussion above that the N_2O microsensor can be used to describe how environmental parameters regulate the net production of N_2O from denitrifying microenvironments. It is well known that nitrification may play an equally important role for production of N_2O by soils (Klemedtsson et al. 1988), but it is likely that the N_2O produced by this process in most instances is produced at very low rates but in large quantities of soil, and such production cannot be studied by sensors where the detection limit, depending on the specific sensor, is about 1 μM.

Transformations of methane

Microsensor for CH_4. A detailed study of methane transformations in methanogenic microenvironments or near oxic-anoxic interfaces has until recently been limited by the unavailability of equipment for measuring methane distribution with a sufficiently high spatial resolution. The introduction of a profiling device based on gas diffusion through a 300 μm wide silicone window in a glass capillary (Rothfuss et al.1994) allowed for the first time relatively detailed depth profiles of methane to be determined. Due to a pronounced dependence of the collection efficiency on the diffusion characteristics of methane in the medium being analyzed, the profiles measured using this technique are, however, probably too inaccurate to allow modelling by diffusion-reaction models and thus the calculation of depth profiles of consumption or production. The technique does, however, have the advantage that all gases can be collected, and the analysis is thus not limited to CH_4. As an example it was possible with a somewhat coarser probe to analyze depth profiles of H_2 in a sediment where the concentrations were in the nanomolar range (Krämer and Conrad 1993), and such analyses would simply not be possible with the electrochemical sensors presently available.

By use of immobilized methanotrophic bacteria it has been possible to make a microscale biosensor for methane (L.R. Damgaard and N.P. Revsbech, manuscript in preparation) which responds linearly to methane over a wide range of concentrations. An example of a calibration curve for such a 30 μm thick sensor is shown in Fig. 4. This specific sensor is not perfectly linear in the full concentration range 0-1 atm. partial pressure of CH_4, but over narrower ranges linearity is sufficient for reliable measurements. Other sensors have shown

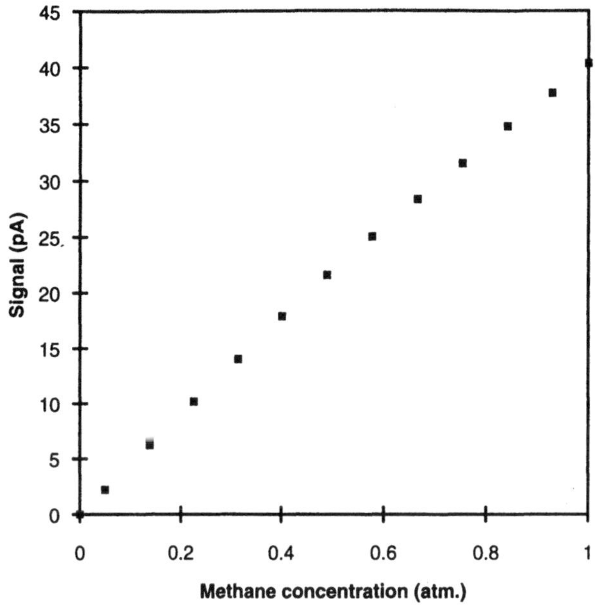

Fig. 4. Calibration curve of a CH_4 microsensor.

perfectly linear calibration curves for the full range of 0-1 atm. partial pressure (Damgaard et al. in preparation). We are unfortunately not able to disclose the design of the sensor as it is currently being considered for patenting. However, the sensor is characterized by a negligible stirring sensitivity, it remains operable for weeks or even months, and except for oxygen, hydrogen sulfide is the only interfering agent. The response for hydrogen sulfide is only about 10% of that for CH_4. The response time is variable depending on the exact design of the sensor, but a 90% response time as short as 10 s can be obtained. The present CH_4 microsensor can only be used under anoxic conditions, which means that methane oxidation cannot be studied. It is expected, however, that the sensor can be modified so that future versions are insensitive to both oxygen and hydrogen sulfide. Such an improved sensor would enable us to perform detailed studies of CH_4 oxidation in a variety of environments.

<u>Methanogenesis in biofilm and sediment.</u> Methanogenesis has been studied in both biofilms and freshwater sediment using the new methane microsensor.

Steady-state concentration profiles of CH_4 in the anoxic layers of a 3-mm thick sewage biofilm submersed in water are shown in Fig. 5. The biofilm had been preincubated for 15 days with 5 mM acetate. After transfer to a medium without any organic substrate, the biofilm still had an endogenous CH_4 production after 1 hour, probably due to decay of biological material accumulated during preincubation (Fig. 5). Adding 10 mM acetate to the water phase significantly stimulated the CH_4 production and the concentration in the deepest analyzed layer increased from 70 to 110 µM (Fig. 5). The production of CH_4 was stimulated by hydrogen as well as by acetate and it was inhibited by sulfate, probably due to competition from sulfate reducers for H_2 and acetate (data not shown, Damgaard et al. in preparation).

An example of a CH_4 profile in a sediment core from an eutrophic lake is shown in Fig. 6. The CH_4 concentration profile exhibited an increase with only a slight upward convex shape from zero at the oxic-anoxic interface at 0.3 mm depth to about 1200 µM at a depth of 8.0 mm. Below this depth the concentration profile gradually approached a value of about 1530 µM. At depths below 8 mm, gas bubbles became visible through the transparent walls of the sampling tubes, and the maximum measured concentrations in these layers actually correspond very well to saturation with pure CH_4 gas (1550 µM at 20 ^{o}C; Yamamoto et al. 1976). A small section of the profile in the depth range of 10-11.5 mm exhibited a constant concentration, most likely reflecting the penetration of a gas bubble by the sensor. The diffusion coefficients of molecules in gases are about 10^4 times higher than the ones in water, so by microsensor analysis gas bubbles always show up as depth intervals with constant concentration (Jørgensen et al. 1983). The slightly higher signal in this gas phase as compared to the adjacent pore water was due to the absence of any external diffusion limitation on the methane consumption by the sensor while the tip was in gas. A similar, although smaller, effect can be observed between turbulent and stagnant liquid, and the

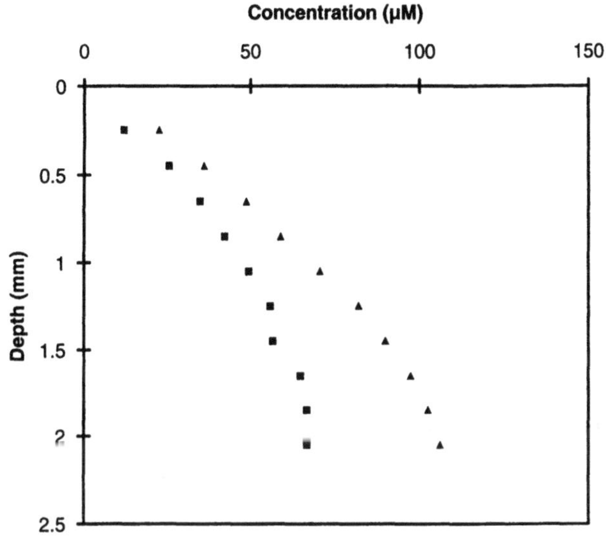

Fig. 5. Methane concentration profiles in a sewage outflow biofilm with (triangular symbols) and without (square symbols) addition of acetate to the overlying water.

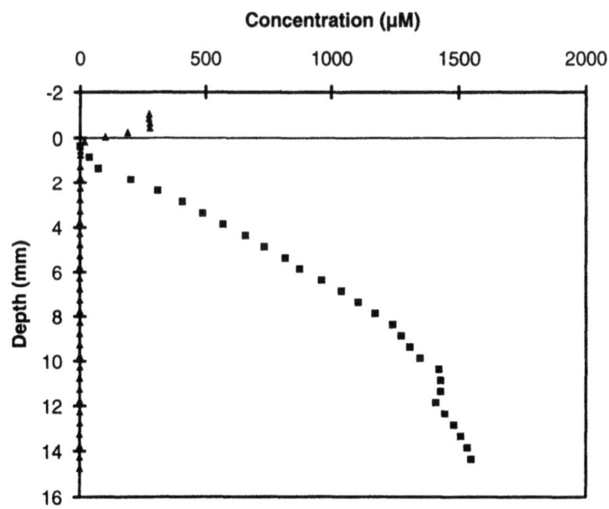

Fig. 6. Methane (square symbols) and oxygen (triangular symbols) concentration profiles in a freshwater sediment.

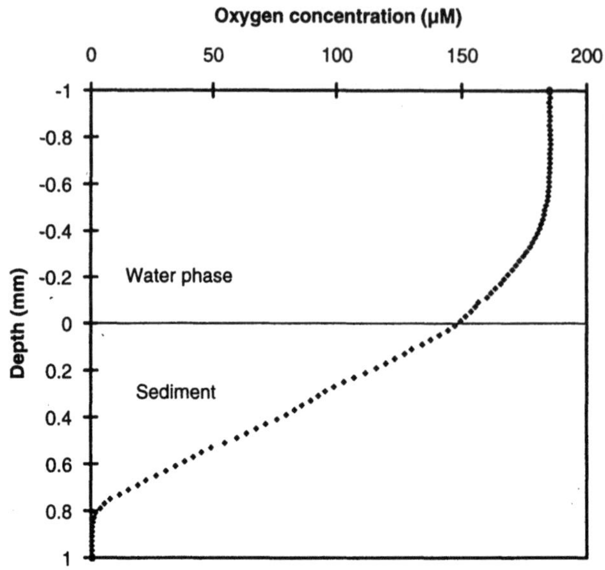

Fig. 7. Oxygen profile in an artificial gradient system with oxygen being supplied from above and methane from below.

phenomenon is therefore often called a "stirring effect". The curvature of the profile below a depth of about 8 mm would usually indicate a high methane production rate, but as the curvature was probably determined by the formation of bubbles, a simple diffusion-reaction model of the methane producing processes cannot be applied on this profile.

Methane oxidation. As mentioned above, the methane microsensor in its present state is sensitive to oxygen, and reliable CH_4 microprofiles in oxic sediments cannot be measured. Methane oxidation can, however, be studied using oxygen microsensors in environments where the major part of the oxygen consumption is due to CH_4 oxidation. Oxygen microsensors have thus been used in CH_4 oxidation studies in natural ecosystems (King et al. 1990). To be able to study CH_4 oxidation in detail we constructed an artificial gradient system with a stirred oxic water phase overlying a layer of freshwater sediment. The sediment was incubated with CH_4 gas supplied from below through a silicone membrane. With this set-up experimental parameters can be carefully controlled, making detailed studies of the behavior of methanotrophic microorganisms possible. Preliminary results show that CH_4 oxidizing bacteria have an ability to concentrate their activity into a very narrow layer where opposing gradients of methane and oxygen meet (Fig. 7). After 40 days of incubation practically all oxygen consumption in the sediment was confined to a 100 µm narrow layer with rates up to 8600 nmol $cm^{-3}h^{-1}$. These measurements of the distribution in time and space of methane oxidation obtained with the oxygen microsensors can now be supported by methane microprofile measurements in the anoxic parts of the sediment. With the projected improvement of the CH_4 microsensor making it insensitive to oxygen, such profiles can be extended into the oxic part of the sediment, yielding more precise information on the dynamics of methane oxidation at a microscale.

Acknowledgements

The development of a microsensor for methane was supported by the EU "Environment" programme, grant EV5V-CT93-0245. The work with a nitrous oxide microsensor was supported by Center for Microbial Ecology under the Danish Biotechnology Programme.

References

Christensen PB, Nielsen LP, Revsbech NP, Sørensen J (1989) Microzonation of denitrification activity in stream sediments as studied with a combined oxygen and nitrous oxide microsensor. Appl Environ Microbiol 55: 1234-1241

Dalsgaard T, Revsbech NP (1992) Regulating factors of denitrification in trickling filter biofilms as measured with the oxygen/nitrous oxide microsensor. FEMS Microbiol. Ecol. 101: 151-164

Damgaard LR, Larsen LH, Revsbech NP. in press. Micro-scale biosensors for environmental monitoring. Trends in Analytical Chemistry

de Beer D, Sweerts J-PRA (1989) Measurement of nitrate gradients with an ionselective microelectrode. Anal Chim Acta 219: 351-356

Firestone MK, Tiedje JM (1979) Temporal change in nitrous oxide and dinitrogen from denitrification following onset of anaerobiosis. Appl. Environ. Microbiol. 38: 673-679

Højberg O, Revsbech NP, Tiedje JM (1994) Denitrification and nitrous oxide accumulation in soil aggregates as measured with a combined microsensor for oxygen and nitrous oxide. Soil Sci. Soc. Am. J. 58: 1691-1698

Jensen K, Revsbech NP, Nielsen LP (1993) Microscale distribution of nitrification activity in sediment determined with a shielded microsensor for nitrate. Appl Environ Microbiol 59(10): 3287-3296

Jørgensen BB, Revsbech NP, Cohen Y (1983) Photosynthesis and structure of benthic microbial mats: Microelectrode and SEM studies of four cyanobacterial communities. Limnol. Oceanogr. 28: 1075-1093

King GM, Roslev P, Skovgaard H (1990) Distribution and rate of methane oxidation in sediments of the Florida Everglades. Appl. Env. Microbiol. 56: 2902-2911

Klemedtsson L, Svensson BH, Rosswall T (1988) A method of selective inhibition to distinguish between nitrification and denitrification as sources of nitrous oxide in soil. Biol. Fert. Soils, 6: 112-119

Krämer H, Conrad R (1993) Measurement of dissolved H_2 concentrations in methanogenic environments with a gas diffusion probe. FEMS Microbiol Ecol 12: 149-158

Revsbech NP (1994) Analysis of microbial mats by use of electrochemical microsensors: Recent advances, p. 135-147. *In* L.J. Stal and P. Caumette (eds.) "Microbial mats. Structure, development, and environmental significance. Springer Verlag

Revsbech NP, Jørgensen BB (1986) Microelectrodes: Their use in microbial ecology. Adv Microb Ecol 9: 293-352

Revsbech NP, Nielsen LP, Christensen PB, Sørensen J (1988) Combined oxygen and nitrous oxide microsensor for denitrification studies. Appl Environ Microbiol 54: 2245-2249

Rothfuss F, Frenzel P, Conrad R (1994) Gas diffusion probe for measurement of CH_4 gradients. In L. J. Stal and P. Caumette Eds.): Microbial Mats. Structure, Development and Environmental Significance. NATO ASI Series. Springer Verlag Berlin Heidelberg, p. 167-172

Yamamoto S, Alcauskas JB, Crozier TE (1976) Solubility of methane in distilled water and seawater. Journal of Chemical and Engineering Data, 21: 78-80

Metabolism of Nitric Oxide in Soil and Soil Microorganisms and Regulation of Flux into the Atmosphere

Ralf Conrad
Max-Planck-Institut für terrestrische Mikrobiologie,
Karl-von-Frisch-Str.
D-35043 Marburg
Germany

Introduction

Nitric oxide plays a central role in the photochemistry of the atmosphere (Crutzen 1979; Logan 1983; Singh 1987). There are several reviews which summarize our knowledge on the emission of NO from soils (Johansson 1989; Conrad, 1990; Davidson 1991; Williams et al. 1992; Meixner 1994). They all agree that emissions from soils are a major source and, with a source strength of about 20 Tg N yr^{-1}, may contribute as much as 40% to the total budget of atmospheric NO. Because of the short atmospheric lifetime most of the emitted NO will soon be deposited again so that the nitrogen loss from one ecosystem may result in unintended fertilization of downwind ecosystems (Conrad 1990; Williams et al. 1992; Meixner 1994).

NO fluxes at the soil-atmosphere interface exhibit a tremendous dynamic. NO emission rates range over four orders of magnitude (Williams et al. 1992; Meixner 1994). Under certain circumstances, soils may even act as a sink for atmospheric NO (Slemr and Seiler 1991; Williams et al. 1992; Conrad 1994). Obviously, the variability of NO fluxes between soil and atmosphere is very high. It is unlikely that intensivation of field campaigns will result in better estimates of the soil source strengths. To arrive at more precise NO source strengths, it seems to be more promising to develop a theoretical understanding of the mechanisms involved in the NO fluxes.

It is obvious that NO turnover in soil is largely the result of microbial processes. The microbial processes that

NATO ASI Series, Vol. I 39
Microbiology of Atmospheric Trace Gases
Edited by J. Colin Murrell and Donovan P. Kelly

may be involved in NO turnover have been reviewed by several authors (Firestone and Davidson 1989; Conrad 1990; Williams et al.1992; Hutchinson and Davidson 1993; Conrad 1995). Metabolism of NO in denitrifying bacteria and in other microorganisms has also been reviewed recently (Ye et al. 1994; Zumft 1993). The reviews concentrate mostly on the aspect of NO production. The aspect of NO consumption is usually neglected.

In the following I will review the aspects of NO production and NO consumption that I feel are important for the exchange of NO between soil and atmosphere. In addition, I will discuss possible approaches to arrive at an understanding of the regulation and a model of the NO flux.

Processes involved in NO production

NO production has been demonstrated in a large number of different microorganisms (e.g. Conrad 1990; Zumft 1993). It has also an important regulatory function in tissues of higher organisms (e.g. Moncada and Higgs 1991; Stamler et al. 1992). The two most important groups of microorganisms seem to be the denitrifiers and the nitrifiers. But there are also other NO-producing microorganisms. All these microorganisms are taxonomically and phylogenetically diverse which means that the metabolic pathways, although they seem to be the same, are not necessarily homologous in the different microorganisms. Despite the large diversity of different microorganisms and metabolic reactions, one can basically distinguish between two types of microbial processes in soil, one involving the reduction of nitrate and the other the oxidation of ammonium. In addition, there are chemical NO production processes.

Chemical NO production

The most important reaction is the chemical decomposition of nitrite (i.e. chemodenitrification) at low pH (Chalk and Smith 1983; Blackmer and Cerrato 1986; Van Cleemput and Baert 1976; McKenney et al. 1990):

$$3\ NO_2^- + 2\ H^+ \rightarrow 2\ NO + NO_3^- + H_2O$$

However, nitrite itself is usually only a minor constituent in soil and is almost exclusively produced by microorganisms (Conrad 1990). Davidson (1991; 1992) proposed that chemical NO production may be important in soil microniches, where locally high nitrite concentrations may arise in water films during freezing or drying, or where H^+ is produced by microbial activity, e.g. by nitrifiers.

Nitrite can be reduced chemically to NO if ferrous iron or other reduced metals are available (Chalk and Smith 1983):

$$NO_2^- + Fe^{2+} + 2H^+ \rightarrow NO + Fe^{3+} + H_2O$$

Organic substrates that allow reduction of ferric iron to ferrous iron and reduction of nitrate to nitrite could also result in production of NO involving both biological (i.e. nitrate reduction and ferric iron reduction) and chemical reactions (i.e. reduction of nitrite to NO). This reaction sequence has been demonstrated in *Escherichia coli* (Brons et al. 1991).

Hydroxylamine is a possible excretion product of nitrifiers (Verstraete and Alexander 1973; Stüven et al. 1992). Nitrite can also be reduced chemically by hydroxylamine (Chalk and Smith 1983). However, the product is usually N_2O:

$$NH_2OH + NO_2^- + H^+ \rightarrow N_2O + H_2O$$

Addition of hydroxylamine to soil indeed resulted only in formation of N_2O but not of NO (Bremner et al. 1980). However, NO was a minor product of this reaction, if catalyzed by preparations of the enzyme nitrite reductase from *Nitrosomonas europaea* (Hooper 1968).

There are also several possible mechanisms that produce NO from the interaction of nitrite with organic compounds, e.g. aromatics (Chalk and Smith 1983; McKenney et al. 1990). In any case, chemical NO production would be limited by the microbial formation of nitrite, so that for practical reasons, the regulation of chemical NO production would have to deal with the regulation of microbial metabolism.

NO production by denitrifiers

Denitrifiers are defined as microorganisms that reduce nitrogen oxides (e.g. nitrate) to gaseous products and couple this reduction to electron transport phosphorylation, i.e. to the generation of energy (Tiedje 1988). The reduction is driven by oxidation of organic substrates or inorganic electron donors. The denitrifiers produce NO as an intermediate during the sequential reduction of nitrate to N_2 (Ye et al. 1994; Fig.1):

$$NO_3^- \text{-->} NO_2^- \text{-->} NO \text{-->} N_2O \text{-->} N_2.$$

The same microorganisms also produce N_2O. The problem is, however, that it is impossible to make a general prediction of how much NO (or N_2O) the typical soil denitrifier would produce. Studies of different species of denitrifiers in pure culture showed large quantitative differences in the rates of NO production which obviously depend on the microbial species and on the physiological conditions (Anderson and Levine 1986; Remde and Conrad 1991a). Even different strains of the same species, e.g. *Pseudomonas aeruginosa*, can display completely different patterns of NO production (Vosswinkel et al. 1991; Kalkowski and Conrad 1991). Studies with soil also showed that the patterns of denitrification products including NO and N_2O were significantly influenced by the type of bacterial flora (Abou-Seada and Ottow 1988; Munch 1989). The reason for this is probably the different regulation of enzyme expression and of enzyme activity in the different denitrifiers. An example is the different regulation patterns exerted by O_2 on the enzymes involved in NO metablism, i.e. the extent of repression and de-repression of nitrate reductase, nitrite reductase, NO reductase and N_2O reductase at different O_2 partial pressures (Remde and Conrad 1991b; Vosswinkel et al. 1991; Bell and Ferguson 1991; Körner 1993; McKenney et al. 1994; Wu et al. 1994). Regulation of enzyme expression was also studied in soil after onset of anaerobiosis (Smith and Tiedje 1979; Firestone and Tiedje 1979; Dendooven and Anderson 1994). The results indicate that while the de-repression of nitrate reductase and nitrite reductase starts within hours, it may take more than one day until the nitrous oxide reductase is synthesized so that N_2O production increases transiently. Production of NO possibly also increases transiently, but this has so far not been studied explicitly.

Some studies suggest that denitrification is possible under fully aerobic conditions (reviewed by Kuenen and Robertson 1988; Lloyd 1993). There are a number of bacteria which are apparently able to denitrify under aerobic conditions, e.g. *Thiosphaera panthotropha* (now *Paracoccus denitrificans*). These bacteria seem to be both denitrifiers and heterotrophic nitrifiers (see below). However, the role of aerobic denitrification in soil, and in particular its role in NO production in soil, is unclear. Experimental evidence indicates that the classical anaerobic denitrification has a much greater potential than aerobic denitrification for NO production in soil. The evidence is based on the fact, that incubation of soil under anaerobic conditions generally stimulates NO production often by orders of magnitude (Johansson and Galbally 1984; Remde and Conrad 1991a; Krämer and Conrad 1990; Remde et al. 1993). However, it is unclear whether this greater potential is actually displayed under in-situ conditions.

Production of NO (and N_2O) was also reported for nitrite-reducing eukaryotic fungi (Burth and Ottow 1983; Shoun et al. 1992; Shoun and Tanimoto 1991). Some of the fungi are also able to produce N_2. *Fusarium oxysporum* contains a copper type nitrite reductase which reduces nitrite to NO (Kobayashi and Shoun 1995). Although these observations suggest that fungi may be denitrifiers, it is still not proven that they can use this type of metabolism for generating energy and supporting growth (Shoun and Tanimoto 1991). Therefore, fungi may not belong to the true denitrifiers *sensu* Tiedje (1988).

NO production by other nitrate reducers

There are microorganisms that reduce nitrate but do not belong to the true denitrifiers, although they produce significant amounts of N_2O and/or NO (Tiedje 1988; Conrad 1990). One group, the nitrate respirers, produce nitrite as a major end product:

$$NO_3^- \dashrightarrow NO_2^-$$

Nevertheless, on a balance to the nitrate dissimilated they can produce as much NO as the denitrifiers (Anderson and Levine 1986; Kalkowski and Conrad 1991).

Another group, the microorganisms with DNRA (dissimilatory reduction of nitrate to ammonium), produce ammonium as major end product (Tiedje 1988):

$$NO_3^- \text{-->} NO_2^- \text{-->} NH_4^+$$

These nitrate reducers, too, can produce substantial amounts of NO (Ji and Hollocher 1988). However, the production mechanism of NO is unclear, since the nitrite reductase of these microorganisms reduces nitrite directly to ammonium without an intermediate (Cole 1988). The DNRA bacteria are physiologically and taxonomically still not very well defined (e.g. Cole 1988; Tiedje 1988; Schumacher et al. 1992). For example, some of the DNRA microbes can also reduce N_2O to N_2 and thus, seem to have some of the enzymes that are typical for denitrifiers (Yoshinari 1980; Teraguchi and Hollocher 1989; Schumacher and Kroneck 1992). A systematic study of NO production in the various DNRA organisms is to my knowledge lacking.

All that these organisms have in common is that the mechanism by which NO is produced is unclear. One mechanism was proposed to be an unspecific reaction of the nitrate reductase with nitrite, thus producing NO and N_2O (Cole 1988; Fig.1). Another explanation could be chemical production of NO from nitrite (Brons et al. 1991). The uncertainty of the production mechanism makes any predictions on the behaviour of these organisms in soil impossible. Experiments with soils indicate, however, that NO production by these organisms may be of environmental relevance (Nägele and Conrad 1990; Blösl and Conrad 1992).

NO production by autotrophic nitrifiers

The autotrophic nitrifiers comprise two physiological groups of bacteria that together oxidize ammonium to nitrate. The bacteria are autotrophic, since they assimilate CO_2 for biomass synthesis. The first group, the so-called *Nitroso*-bacteria, oxidizes ammonium to nitrite. They obligately

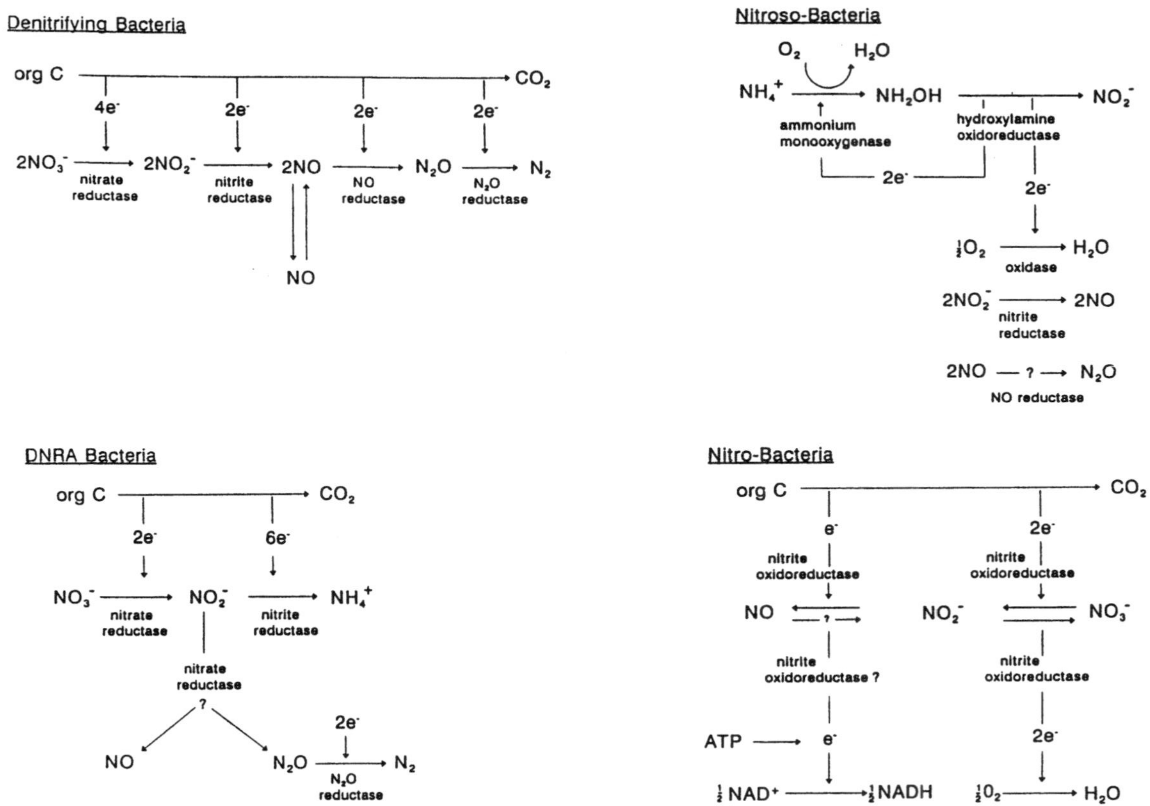

Fig. 1. Schemes (mostly hypothetical) of NO production and NO consumption in denitrifiers, in bacteria with dissimilatory reduction of nitrate to ammonia (DNRA), in ammonium-oxidizing nitrifiers (Nitroso-bacteria), and in mixotrophically growing nitrite-oxidizing nitrifiers (Nitro-bacteria).

require O_2 for the initial reaction that is catalyzed by an ammonium monooxygenase (Hooper 1989):

$$NH_4^+ \text{ --> } NH_2OH \text{ --> } NO_2^-$$

The *Nitroso*-bacteria produce significant amounts of NO (Lipschultz et al. 1981; Anderson and Levine 1986), obviously by the reduction of the produced nitrite (Remde and Conrad 1990; Fig.1). The reduction of nitrite is possibly coupled to energy generation (Abeliovich and Vonshak 1992).

Stüven et al. (1992), on the other hand, suggested that the produced nitrite is chemically reduced to NO by NH_2OH. Indeed, the authors found low amounts of NH_2OH being excreted into the medium, when the bacteria grew mixotrophically with formate plus ammonium as substrate, and also observed NO production. However, the authors performed no sterilization experiments so that NO production may also be due to metabolic reduction of nitrite to NO with NH_2OH as physiological electron donor, a reaction that was first described by Hooper (1968).

The *Nitroso*-bacteria also produce N_2O. However, it is still not known if the production mechanism is due to a sequential pathway (NO_2^- --> NO --> N_2O --> N_2) as in denitrifiers. Production of N_2 is sometimes but not always observed in *Nitroso*-bacteria (Poth 1986).

The NO/N_2O ratios that were so far observed vary by three orders of magnitude (Anderson and Levine 1986; Remde and Conrad 1990; Stüven et al. 1992). One reason for the large variation may be the fact that NO and N_2O are only minor products which constitute not more than a few percent of the nitrite produced and that their production rates are regulated by environmental factors, such as the availability of O_2. Maximum production of N_2O and, to a lesser extent, NO is reached at low O_2 (Lipschultz et al. 1981; Remde and Conrad 1990; Anderson and Levine 1986; Anderson et al. 1993). This is reasonable, since the bacteria must save the little available O_2 for the ammonium monooxygenase reaction, and thus better replace O_2 by nitrite for the oxidation of hydroxylamine (Fig.1). The reason why NO is also produced under oxic conditions at relatively high rates may be due to the necessity to regulate the pH (Groeneweg et al. 1994). The reduction of nitrite to NO in the periplasm of the bacteria consumes 2 protons and thus favours the deprotonation of NH_4^+

to NH_3, a prerequisite for transport and oxidation of ammonia at neutral pH.

Another reason for the variations in NO/N_2O ratios may be that the different strains used by the different investigators have different patterns of regulation. Such differences are not unlikely, since the autotrophic nitrifiers are not a phylogenetically homogenous group (Head et al. 1993; Orso et al. 1994; Teske et al. 1994).

The second group of nitrifiers, the so-called *Nitro*-bacteria, oxidize nitrite to nitrate with O_2 as electron acceptor (Hooper 1989; Bock et al. 1991):

$$NO_2^- \text{-->} NO_3^-$$

Alternatively, under anoxic conditions, the *Nitro*-bacteria can also reduce nitrate to nitrite (Freitag et al. 1987), or nitrite to NO (Ahlers et al. 1990), if they have a suitable electron donor, e.g. pyruvate. All these reactions are apparently catalyzed by the same enzyme, nitrite oxidoreductase (Bock et al. 1991; Fig.1). Mixotrophically growing cultures produce NO at 5-8% air saturation (Bock et al. 1991). NO is postulated to be an important intermediate in energy generation (Freitag and Bock 1990; Bock et al. 1991). However, an unequivocal theory for the metabolic role of NO and systematic measurements of NO production in *Nitro*-bacteria are so far lacking.

NO production by heterotrophic nitrifiers

The heterotrophic nitrifiers are defined as heterotrophic bacteria that are able to nitrify in the sense of a "biological transformation of nitrogen in organic and inorganic compounds from a reduced to a more oxidized state" (Alexander et al. 1960). The heterotrophic nitrifiers comprise bacteria and fungi. However, the biochemical pathway of heterotrophic nitrification is still not known. There may be many different pathways depending on the type of organism. The reason why heterotrophic bacteria nitrify is also unknown (Focht and Verstraete 1970; Killham 1986; Kuenen and Robertson 1988). For practical purposes, one might define heterotrophic nitrification as nitrate production that is not due to autotrophic nitrifiers. In principle, two different pathways are possible, an inorganic and an organic pathway (Killham 1986):

inorganic: NH_4^+ --> NH_2OH --> NO_2^- --> NO_3^-

organic: $R\text{-}NH_2$ --> R-NHOH --> R-NO --> $R\text{-}NO_2$ --> NO_3^-

It is possible that organic nitrogen compounds are involved in only some steps of the oxidation pathway. For example, nitrosation reactions were involved in nitrification of ammonium in soil (Verhaegen et al. 1988). The fate of these organic nitrogen compounds and their potential role in NO metabolism in soil is unknown.

Interestingly, many of the heterotrophic nitrifiers are also able to denitrify (Castignetti and Hollocher 1991) and many denitrifiers are able to heterotrophically nitrify (Castignetti and Hollocher 1984). Many of these heterotrophic nitrifier-denitrifiers seem to denitrify under aerobic conditions (Kuenen and Robertson 1994). *Alcaligenes faecalis* is one of these heterotrophic nitrifier-denitrifiers (VanNiel et al. 1992), and this species was also shown to produce substantial amounts of NO and N_2O from ammonia under oxic conditions (Papen et al. 1989; Anderson et al. 1993).

In some forest soils, NO and N_2O are possibly also produced by heterotrophic nitrification (Papen et al. 1993). However, the distinction of autotrophic and heterotrophic nitrification is not unambigous. Heterotrophic nitrification is usually quantified as that part of nitrification that is insensitive to inhibition by acetylene or nitrapyrin. These two compounds (and others) inhibit the ammonium monooxygenase reaction of autotrophic nitrifiers (Bedard and Knowles 1989). It is obvious that heterotrophic nitrifiers that oxidize organic nitrogen without the intermediate production of ammonium, and thus do not require ammonium monooxygenase activity, are also not susceptible to inhibition by nitrapyrin (Barraclough and Puri 1995). If nitrification is not inhibited by acetylene or nitrapyrin, it may be relatively safe to conclude that it is due to heterotrophic nitrifiers (Schimel et al. 1984; Kreitinger et al. 1985; Duggin et al. 1991). Perhaps, one may even conclude that heterotrophic nitrification operates in these cases on organic nitrogen rather than on free ammonium.

However, a more careful interpretation is required if inhibition is observed (Papen et al. 1993). It is possible that the inhibited nitrification is due to heterotrophic nitrifiers that oxidize ammonium instead of organic nitrogen. Several studies show that the heterotrophic nitrification by *A. faecalis* and *T. pantotropha* is also inhibited to some extent by nitrification inhibitors (Robertson et al. 1989; Papen et al. 1993; Anderson et al. 1993). Possibly these bacteria activate ammonium also with an ammonium monooxygenase. An alternative reaction that does not involve O_2, e.g. conversion of ammonia to hydroxylamine by a dehydrogenase, is hardly conceivable since the redox potential of NH_4^+/NH_2OH is very high (E_o' = +906 mV). However, evidence is accumulating that oxidation of ammonium can be coupled to the reduction of nitrate (VanderGraaf et al. 1995; Mulder et al. 1995). *Nitrosomonas eutropha* was reported to couple the oxidation of ammonium to the reduction of nitrite in the absence of O_2 (Bock et al. 1995). However, the mechanism of ammonium activation and its possible susceptibility to inhibitors awaits elucidation.

Processes involved in NO consumption

NO is a radical with an unpaired electron in the π* orbital. Therefore, it is highly reactive and can be decomposed by several chemical and biological reactions. The redox potentials for both reduction of NO to N_2O (E_o' = +1173 mV) and oxidation of NO to NO_2 (E_o' = +638 mV) are relatively high (Conrad 1990). Therefore, NO is easily reduced under physiological conditions, e.g. coupling to the oxidation of NADH to NAD^+. However, it can only be oxidized with reagents having a relatively positive redox potential. Thus, coupling to the reduction of O_2 to H_2O is possible, but not to the reduction of NO_3^- to NO_2^- (E_o´ = +432 mV) or NO_2^- to NH_4^+ (E_o´ = +340 mV). Other redox potentials involved in nitrogen cycling are found in Conrad (1990).

Chemical NO consumption

NO is unstable in air at room temperature, since it reacts chemically with O_2:

$$2\ NO + O_2 \rightarrow 2\ NO_2$$

However, this reaction takes only place if the NO concentration is larger than about 10 ppmv, since two molecules of NO have to combine with one molecule of O_2. At the low ambient concentrations of NO (<10 ppbv), such an event is so rare that the half life of NO is >100 h (Galbally and Roy 1978).

The same termolecular reaction takes place in aqueous solution (Kharitonov et al. 1994). The half life of NO at concentrations of 100 nM NO (equivalent to about 55 ppmv NO in the gas phase) and 230 μM O_2 was calculated to be 2 h (Kharitonov et al. 1994). In aqueous solution, the NO_2 produced obviously reacts further to produce nitrite as sole reaction product (Ignarro et al. 1993). The hypothetical reaction sequence is:

$$2\ NO_2 + 2\ NO \rightarrow 2\ N_2O_3$$

$$2\ N_2O_3 + 2\ H_2O \rightarrow 4\ NO_2^- + 4\ H^+$$

In the atmosphere, NO reacts rapidly with O_3 to NO_2; this reaction takes also place at the trace concentrations of NO in the atmosphere (Logan 1983):

$$NO + O_3 \rightarrow NO_2 + O_2$$

Since O_3 is rapidly depleted, this reaction only plays a role in sunlight (Logan 1983). In field experiments, the possible interference with O_3 is often avoided by using static boxes, where O_3 is rapidly depleted, or dynamic boxes with purified air, or by using other corrections. The advantages and disadvantages of the different techniques are discussed by Williams et al. (1992) and Meixner (1994).

NO also reacts with all kinds of other radicals, e.g. organic peroxides. In atmospheric chemistry, the potential reactions are relatively well studied (Singh 1987; Williams et al. 1992). One reaction product, for example, peroxyacetyl nitrate (PAN) is an important reservoir species that releases NO at increased temperatures or by photochemical reactions and thus allows the long-range transport of NO.

In the tissues of higher organisms, where NO is produced as a signal transducer, NO seems to be rapidly oxidized to nitrate by reactions with superoxide and/or hemoproteins (Stamler et al. 1992; Ignarro et al. 1993), e.g.:

$$NO + O_2^- \longrightarrow NO_3^-$$

Similar reactions are likely in microorganisms where the decomposition of high NO concentrations is necessary to avoid toxic effects (Lipton et al. 1993; Zumft, 1993). NO also reacts with organic peroxides present in the cell (Padmaja and Huie 1993).

In soil, however, chemical oxidation reactions probably play only a marginal role, since they are strongly dependent on the concentrations of the reactants. In contrast to tissue cultures, where cellular NO concentrations can reach values of >400 nM (equivalent to about >220 ppmv NO) (Malinski and Taha 1992), the soil atmosphere usually exhibits concentrations well below 1 ppmv NO (Rudolph and Conrad 1995). Chemical reactions are then accordingly slower. Indeed, NO consumption in soil was usually inhibited by sterilization (e.g. Remde and Conrad 1991a) or was orders of magnitude lower compared to the uninhibited control (Koschorreck and Conrad, unpublished).

NO consumption by denitrifiers

Since NO is an intermediate in the sequential reduction pathway of nitrate to N_2, it is plausible that denitrifiers can utilize NO as electron acceptor (Fig.1). Reduction of NO to N_2O is catalyzed by the NO reductase and is coupled to electron transport phosphorylation (Zumft 1993; Ye et al. 1994). The affinity of denitrifiers to NO is very high; both pure cultures and denitrifying soil samples exhibit NO consumption kinetics with K_m values < 8 nM (Remde and Conrad 1991a; Baumgärtner and Conrad 1992a; Schuster and Conrad 1992; Schäfer and Conrad 1993; McKenney et al. 1994). For many denitrifiers it seems that once the NO reductase is synthesized, the enzyme is relatively insensitive to O_2, so that NO consumption by denitrification can take place even in well-aerated soil (Remde and Conrad 1991a,b; Schäfer and Conrad 1993).

NO consumption by other microorganisms

Comparatively little is known about the NO consumption by other microorganisms, although such an activity has been demonstrated in *Nitrobacter* sp. (Freitag and Bock 1990); in heterotrophically nitrifying methanotrophic bacteria (Krämer et al. 1990); in *Alcaligenes faecalis* (Anderson et al. 1993); and in a heterotrophic *Pseudomonas* sp., strain PS88 (Baumgärtner 1991; Koschorreck and Conrad, in prep.).

The NO consumption in *Nitro*-bacteria is postulated to be the result of a peculiar mechanism for the generation of energy and reducing equivalents (Bock et al. 1991; Fig.1). The oxidation of NO to nitrite seems to be coupled to the generation of NADH in a reaction that requires the input of energy (Freitag and Bock 1990; Bock et al. 1991). However, NO metabolism in this bacterial group needs further research.

More research is also required for the heterotrophic nitrifiers. In particular, it is unclear if NO is really consumed by reduction to N_2O as suggested by the frequent occurrence of denitrifying capacities in heterotrophic nitrifiers (see above). Soils exhibited increased rates of NO consumption if they were enriched in methanotrophic bacteria, suggesting that this group of heterotrophic nitrifiers may contribute significantly to NO consumption in soils (Krämer et al. 1990; Bender and Conrad 1994).

The NO consumption process in the heterotrophic *Pseudomonas* strain PS88 was recently characterized in our laboratory (Koschorreck and Conrad, in prep.). This strain had been isolated as one of the dominant species in a calcic cambisol that exhibited increased NO consumption under oxic versus anoxic conditions (Baumgärtner 1991). Strain PS88 is neither a nitrifier nor a denitrifier, and its growth is not stimulated by NO. However, it is able to oxidize NO in the presence of O_2 stoichiometrically to nitrate as the only detectable product. The reaction depends on living cells and on the presence of O_2. More than 80% of the activity is lost when the cells are disintegrated by ultrasonication or French press. However, some activity is still detectable in cell extracts (mainly in the particulate fraction). Possibly, the oxidation of NO is a chemical reaction with superoxide (or other radicals) that itself is generated during aerobic microbial metabolism.

It is noteworthy that the kinetics of all these NO-consuming microorganisms are distinct from those of denitrifiers (Conrad 1995). Whereas NO consumption by denitrifiers is saturated at low NO concentrations (K_m < 8 nM), that by the other microorganisms is not saturated at NO concentrations >30 nM. Whereas NO consumption by denitrifiers is stimulated by anoxic conditions, that by the heterotrophic *Pseudomonas* sp. is stimulated by oxic conditions.

Microbial metabolism of NO in soil

The large number of different processes and microorganisms which are potentially involved in either NO production or NO consumption make it difficult to predict the flux of NO between soil and atmosphere. It should be emphasized that the processes involved in NO turnover are a composite of different species of microorganisms. Thus, NO is not simply produced by, for example, the denitrification, but by the different denitrifying bacteria in soil. These denitrifying bacteria are not necessarily homogenous. Quite in contrast, they belong to many different species which are mostly phylogenetically and taxonomically unrelated (see above). The same is true for nitrifying microorganisms. It appears, that the regulation of the NO turnover process in the various bacterial species is not identical even if they belong to one and the same physiological group, e.g. the denitrifiers (see above). Therefore, to understand NO metabolism in soil on a genuinely mechanistic basis, microbial ecologists would like to determine which species of NO-metabolizing microorganisms are present in a particular soil.

It is possible that NO-producing microbial populations exist that are specifically adapted to the properties of a particular soil, e.g. to the soil temperature, pH, water potential and nutrient status. Thus, nitrate-reducing populations that are adapted to pH (Parkin et al. 1985; Blösl and Conrad 1992) or to temperature (Saad and Conrad 1993) have been demonstrated. It was longtime believed that nitrification in acidic soil is exclusively due to heterotrophic nitrifiers (Lang and Jagnow 1986; Stroo et al. 1986; Kuenen and Robertson 1988). Later, however, autotrophic nitrifiers were also shown to be active in acidic soils (DeBoer et al. 1989; Martikainen and DeBoer 1993) and could be adapted to low pH (DeBoer et al. 1991; Allison and Prosser 1993; Hayatsu 1993).

At the moment, we are far from understanding the microbial diversity of soils and how the populations of the various types of microorganisms are adapted to particular soil conditions. There are some promising approaches comprising molecular techniques. Thus, first attempts to characterize the soil microbial populations by detecting the genes that code for a particular enzyme involved in NO metabolism, e.g. the nitrite reductase, were quite successful (Smith and Tiedje 1992; LinnevonBerg and Bothe 1992). This approach may even allow to distinguish between different types of enzymes, e.g. the heme *cd* or nonheme copper type of nitrite reductase (Coyne et al.1989). However, it will still be a long way until it is possible to characterize NO metabolism in soil from the knowledge of the resident microbial flora.

In the meantime, our theoretical understanding of the NO metabolism in soil will be based on the measurement of bulk activities in different soils and under different soil conditions. This understanding will be assisted by using conceptual models.

Control of NO flux

A mechanistic basis for the understanding of how NO fluxes come about and how they are controlled by the environment is highly desirable. Even though it appears almost impossible to develop a mechanistic model that includes the actual controls of the nitrogen transformation reactions on the level of the microorganisms, a theoretical background may help for devising simpler models that consider as many of the actual controls as feasible.

A conceptual model illustrating the different levels at which process control may be studied is given in Fig.2.

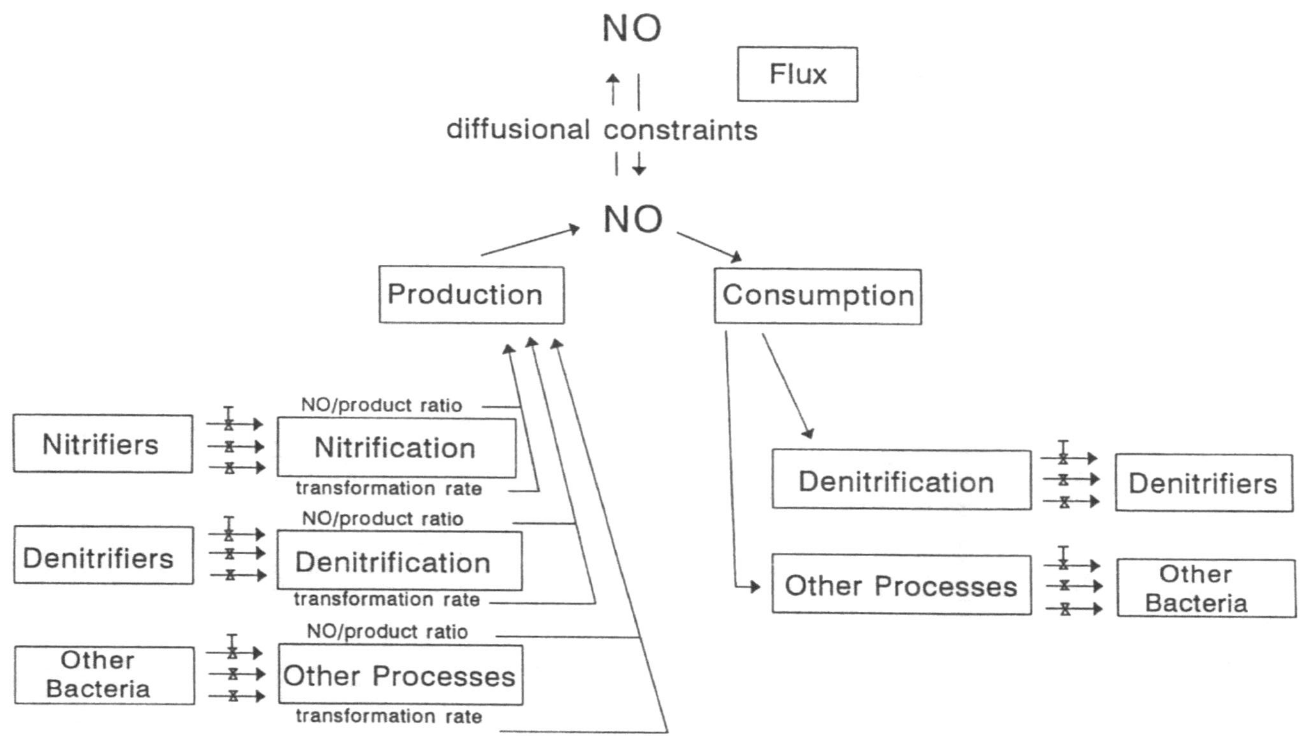

Fig.2. Conceptual scheme of the different levels at which the control of NO flux by soil variables can be studied. The valves indicate the actual sites where NO metabolism is under environmental control by a hierarchy of controlling variables (exemplified in Fig.3, but potentially different for each organism).

Accordingly, regulation may be studied at the levels of (1) net flux of NO, (2) total production and consumption of NO, (3) nitrogen transformation rates and NO/product ratios, (4) processes contributing to NO production and NO consumption, and (5) microorganisms involved in the individual processes. The controls at the higher levels integrate over those at the lower levels. The sum of all microbial activities ultimately results in the NO flux. The scheme in Fig.2 also illustrates the different stages of complexity of modelling approaches. Regulation models on the level of the NO flux would be largely empirical. However, models would become more and more mechanistic, and thus would offer an increasingly better theoretical understanding, if they included the lower levels of control, and differentiated between control of NO production and NO consumption, differentiated beween different NO production and consumption processes, or even differentiated between different microbial soil floras.

Control of NO flux by production, consumption and diffusion of NO

The flux of NO at the soil surface is the result of NO production, of NO consumption and of NO diffusion. Knowing these variables should allow the calculation of the NO flux by a relatively simple diffusion model (Galbally and Johansson 1989). Recently, we were able to confirm the principal applicability of such a model in controlled laboratory experiments on intact soil cores (Rudolph and Conrad 1995; Rudolph et al. 1995). The results confirm that NO consumption is as important for NO flux as NO production resulting in a compensation concentration which determines the direction and the magnitude of the flux (Conrad 1994). The NO flux at the surface can be modelled either from the vertical NO profiles in soil or from NO turnover rates determined in soil samples. Laboratory experiments of turnover rates plus knowledge of diffusion characteristics of soils also predicted NO fluxes which could be verified in field experiments within a factor of 2-4 (Remde et al. 1993). Hence, it appears possible to determine NO fluxes in the field by laboratory measurements which are logistically less demanding than field measurements. However, the calculation of the NO flux still requires the measurement of NO production and consumption rates under different environmental conditions. The next straightforward step would be to predict NO production and consumption from variables that are easier to determine.

Control of NO production by nitrogen transformation rates

It is obvious that NO production is controlled by the same environmental variables that also control the turnover of nitrogen in soil. Conceptual models of considerable complexity have been designed to outline the hierarchy of the more proximal and more distal variables that control the nitrogen flow through denitrification and nitrification (Tiedje 1988; Robertson 1989; Fig.3). The proximal regulators include ammonium and O_2 for nitrification, and O_2, nitrate and organic carbon for denitrification. Although the availability of ammonium for nitrification and of O_2 for denitrification are considered to be the two most important proximal regulators, it cannot be excluded that in some soils other regulators are more important. This will especially be the case if the oxidation of nitrogen is not simply due to autotrophic nitrification, but to one of the less well understood heterotrophic nitrification reactions, or if the reduction of nitrogen is not due to true denitrifiers, but to DNRA organisms, heterotrophic nitrifier-denitrifiers or a tight coupling between oxidation-reduction reactions (Rysgaard et al. 1993; Nielsen and Revsbech 1994; Risgaard-Petersen et al. 1994). In addition, soil temperature and pH are variables that can be expected to affect every process and thus are most probably also important regulators for the nitrogen fluxes. These variables also affect the composition of the relevant microflora (e.g. Parkin et al. 1985; Blösl and Conrad 1992; Saad and Conrad 1993a) and thus may affect the pathways and the hierarchy of controls of nitrogen transformation.

It should also be emphasized, that it is not trivial at all to measure the gross rates of ammonium oxidation or nitrate reduction in soil. This is even more the case, if the

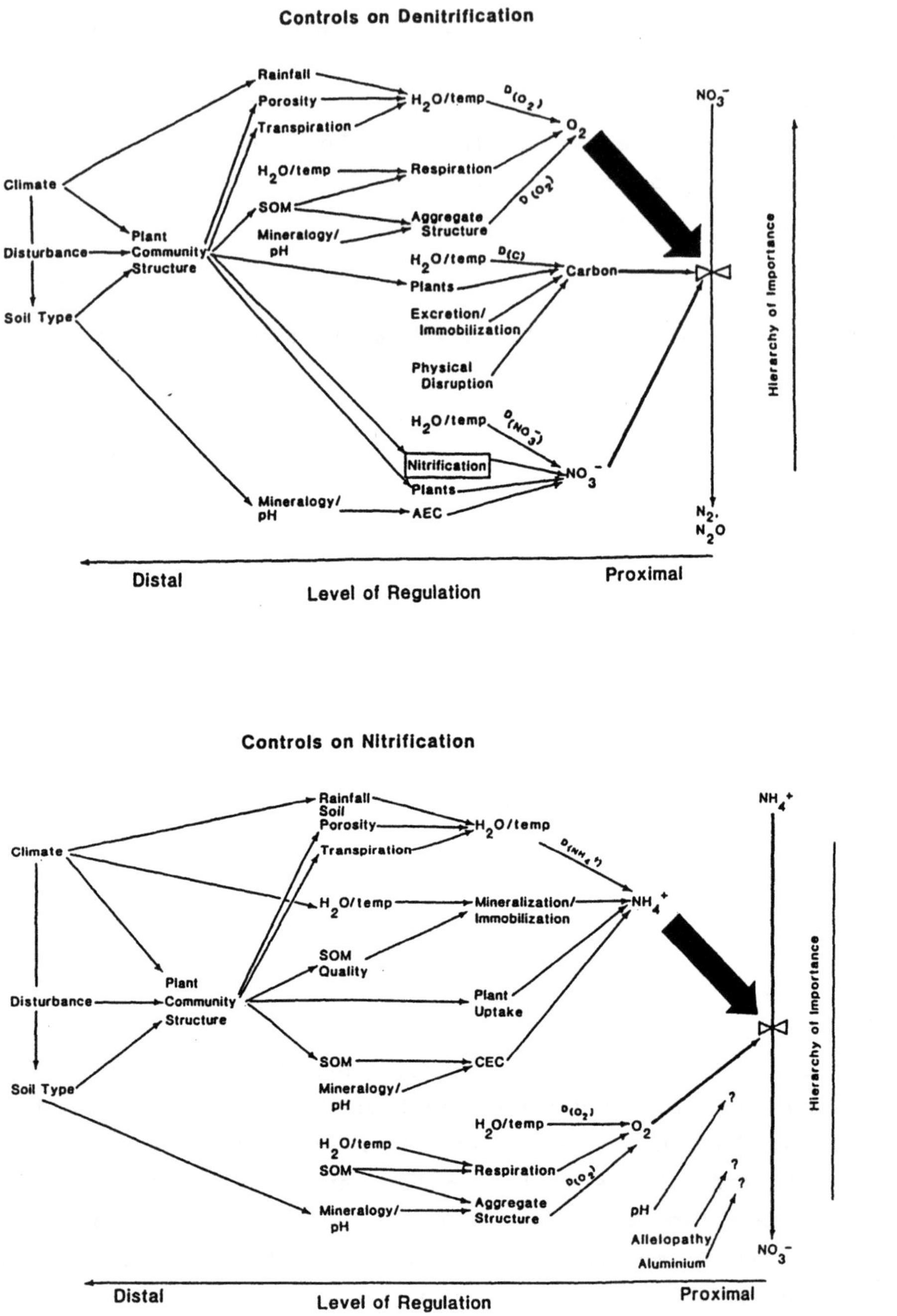

Fig.3. Conceptual scheme of the hierarchy of the major controllers of nitrification and denitrification in soil (adapted from Robertson 1989; and Hutchinson and Davidson 1993).

oxidation or reduction reactions are composites of different processes (e.g., autotrophic and heterotrophic nitrification; denitrification and DNRA) which are presumably controlled by different variables. The techniques to measure gross rates of nitrogen transformation usually involve ^{15}N labelling and acetylene inhibition. Some recent examples of studies in soils are found in Davidson et al (1991), Tietema and Wessel (1992), Hart et al. (1994), Groffman et al. (1993), Pennock et al. (1992), Aulakh et al. (1992). However, denitrification rates determined with the acetylene inhibition technique can be underestimated if denitrification is tightly coupled to nitrification (Rysgaard et al. 1993; Nielsen and Revsbech 1994; Risgaard-Petersen et al. 1994). Labelling studies with ^{15}N require the addition of nitrogen together with the label, thus possibly affecting the nitrogen transformation rates.

In addition, it is also not trivial to measure the soil variables that potentially exert a controlling effect on nitrogen transformation. Concentration measurements of nutrients, e.g. ammonium, can be misleading due to soil heterogeneity (Davidson and Hackler 1994). The availability of O_2 at the site of the nitrogen-transforming bacteria must either be modelled from the soil diffusional characteristics and the respiration rates (e.g. Renault et al. 1994) or must be assessed for each particular microniche (Sexstone et al. 1985; Højberg et al. 1994). On a global basis, our systematic knowledge of the controlling soil variables is still too poor to make generalisations about hierarchies.

Control of NO production by the NO/product ratio

The control of NO production is further complicated by the fact that NO is only a minor product of the different NO transformation processes. Another conceptual model, the "hole-inthe-pipe-model" accounts for this complication (Firestone and Davidson 1989; Fig.4). This model assumes that the rate of NO production (or N_2O production) is given by the rate of nitrogen transformation (the pipe) and the ratio of NO to product (the hole). Both the transformation rates and the NO/product ratios are under the control of environmental

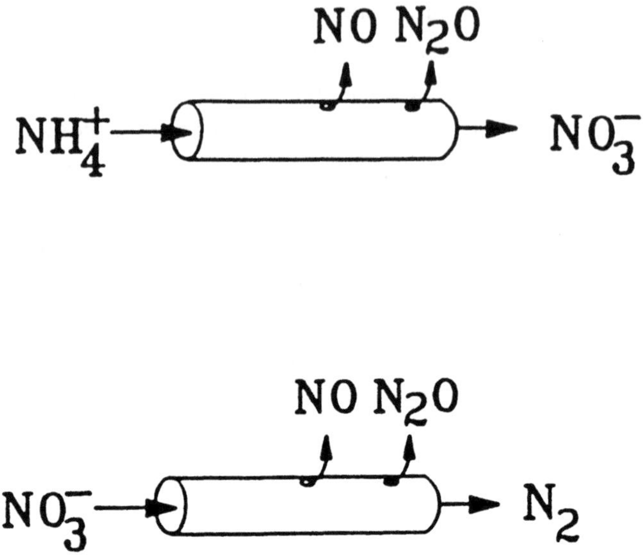

Fig.4. "Hole-in-the-pipe" model describing the two levels of control of NO production during nitrification and denitrification: the nitrogen transformation rates are represented by the flow through the "pipes" and the NO/product ratios (or N_2O/product ratios) by the size of the "holes" (adapted from Firestone and Davidson 1989).

variables as outlined above. This means, that in addition to the controlling mechanisms of the NO/product ratio the transformation rate should be known. Basically, the control of the product ratio is due to the control of the enzymes upstream and downstream of NO in the nitrogen transformation pathway of a particular microorganism. The steady state concentration of NO (or N_2O) can be modelled from the kinetics of the producing and consuming enzymes. In fact, this has quite successfully been done for denitrification both in culture (Betlach and Tiedje 1981; Zafirou et al. 1989) and soil (McKenney et al.1984) or activated sludge (VonSchulthess et al. 1995). However, systematic quantitative studies of the control of NO/product ratios by environmental variables in microbial cultures and in soils are still rare. In soil, the problem is compounded by the fact, that NO production usually is the result of more than one microbial species and, even worse, more than one type of process. Since nitrification, denitrification and other reactions may contribute to nitrogen transformation, one has to make sure to relate the right amount of NO production to the respective process.

Usually, NO production by nitrification seems to account only for a few percent of total nitrogen transformation in soil (e.g. Remde and Conrad 1991c; Baumgärtner and Conrad 1992b; Hutchinson et al. 1993). NO production from denitrification made up a similarly low percentage when measured in oxic soil, but made up >10% when measured in completely anoxic soil (Remde and Conrad 1991c; Baumgärtner and Conrad 1992a; Nägele and Conrad 1990). The percentage contribution of NO to nitrate reduction increased with decreasing pH of the soil (Nägele and Conrad 1990). More systematic studies on the regulation of NO/product ratios by environmental variables are needed.

It should be noted, that the NO/product ratios may be different, if the net release of NO instead of the gross production of NO is measured, as done for example in field measurements (Shepherd et al. 1991; Hutchinson and Brams 1992; Davidson et al. 1993). The net release of NO is a composite of NO production and NO consumption, which are independently regulated by soil variables (see below).

Contribution of different processes to NO production

The "hole-in-the-pipe-model" distinguishes between nitrification and denitrification, i.e. two different "pipes" and two different "holes". It is therefore necessary to quantify the NO/product ratios separately for the two pathways. Besides measuring the gross rates of nitrification and denitrification (see above), it must be identified how much of the NO is produced by each of these two pathways. The analysis is again complicated by the fact that heterotrophic nitrification and nitrate reduction processes other than denitrification can be involved.

Conducting ^{15}N-labelling as well as specific stimulation and inhibition experiments, Remde and Conrad (1991c) showed that either nitrification or denitrification can be the more important pathway of NO production. The most practical approach for routine determinations is the application of a suitable nitrification inhibitor (nitrapyrin, acetylene, dicyandiamide, etc.) to determine the percentage contribution of nitrification to NO production. The percentage contribution can then be assessed as a function of soil variables (Tortoso and Hutchinson 1990; Schuster and Conrad 1992; Davidson 1992; Davidson et al. 1993; Hutchinson et al. 1993; Papen et al. 1993; Paul et al. 1993; Skiba et al. 1993). The data base is not satisfactory, but most of the studies indicate that soil moisture, for example, has a relatively low impact on the pathway (nitrification versus denitrification) by which NO is produced. This finding is in contrast to N_2O production which is mainly due to nitrification at soil moisture levels below field capacity and mainly due to denitrification at soil moisture levels above field capacity (Davidson 1993). As a consequence, the ratio of N_2O/NO produced shows a significant positive correlation with soil moisture, although with a large scatter of the individual data so that soil moisture explains less than 32% of the total variability (Keller and Reiners 1994).

It should be noted, that the studies applying nitrification inhibitors only allow the determination of the relative contribution of nitrification to total NO production. The relative contribution of denitrification is then determined by difference to total NO production. This contribution may be overestimated if processes are involved that are not completely inhibited by the nitrification inhibitor (e.g. heterotrophic nitrification; see above) and may be underestimated if denitrification is tightly coupled to nitrification (see above).

It should also be noted that the determination of NO/product ratios for the "hole-in-the-pipe" model requires the simultaneous determination of rates of nitrification and denitrification. There are no systematic studies of the control of NO/product ratios by soil variables. Therefore, it is presently not possible to link the controls of NO production to the more general models of hierarchical control of nitrogen transformation. Of course, such a linkage would be desirable for a general theory of control of nitrogen transformation. From a practical point of view, however, a hierarchical control scheme for NO production alone may be sufficient. Since gross rates of NO production are much easier and much more accurate to measure than gross rates of nitrification and denitrification, progress could be made more rapidly. This approach seems to be especially reasonable, since our theoretical understanding of the hierarchical controls of the transformation rates of total nitrogen is too small (at least on a global scale) to warrant sticking to it as the basis for modelling. Of course, we will lose the information of how much the control affects the flow through the "pipe" versus the diameter of the "holes".

Control of NO consumption

The flux of NO is not only influenced by the NO production rate, but even more by the NO consumption rate. In fact the diffusion model of Galbally and Johansson (1989) is more sensitive to changes in consumption than in production of NO (Remde et al. 1993). Indeed, production and consumption of NO seem to be independently regulated by soil variables, such as O_2 availability or temperature (e.g. Krämer and Conrad 1991; Saad and Conrad 1993b). NO consumption in soil changed over the season as dynamically as NO production (Baumgärtner and Conrad 1992b). In the two soils studied, >50% of the variability (temperature kept constant) was explained by the negative correlation of NO consumption with soil moisture (Baumgärtner and Conrad 1992b).

We know little about what processes beside denitrification are involved in NO consumption (see above), and we know even less about the regulation of NO consumption by environmental variables. NO consumption by denitrification seems to be stimulated by organic carbon also by nitrate (Remde and Conrad 1991c; Baumgärtner and Conrad 1992a; Schuster and Conrad

1992). However, increasing soil moisture showed no significant influence on NO consumption by denitrification, whereas it inhibited NO consumption by other processes (Schuster and Conrad 1992).

It is presently not possible to quantitatively differentiate between the contribution of denitrification versus other processes to NO consumption. Qualitatively, however, there are two significant differences (Conrad 1995a): (1) the K_m of NO consumption by denitrification is much lower than that by other processes; (2) NO consumption by denitrification is enhanced under anoxic conditions, whereas that by other processes is inhibited. More research is necessary to allow a more detailed differentiation of the NO consumption processes and to construct hierarchical control schemes.

Control of NO flux - empirical approaches

Empirical approaches assess the controls on NO flux in a direct way, e.g., Williams et al. (1992) predicted NO fluxes from a simple exponential temperature function that was supplemented with factors that were chosen as typical for particular ecosystems. The advantage of such an approach is that it is not necessary to differentiate between production and consumption of NO, or between the different processes involved in production or consumption, since all processes are included in one black box. Exactly this, on the other hand, is the disadvantage. The disadvantage is especially severe, if the variabilities in the field are so high, that no generalizations can be made and a new black box has to be created for each field site. However, compromises are possible by considering some of the lower levels of regulation (Fig.2). For example, it is possible to assess separately the controls on the NO production by nitrification and on the NO production by other processes as outlined above. Hopefully this will allow some generalizations and a reduction of the number of black boxes necessary for modelling NO fluxes on a global scale.

For example, recent measurements of the relative contribution of nitrification to NO and N_2O production in different soils (n=8) at standard temperature (25°C) and standard soil moisture (60% whc) indicate that more than 70% of the remaining variability can be explained by soil texture with nitrification dominating in sand versus denitrification dominating in clay (Bollmann and Conrad, unpublished). Also, >70% of the variability in total NO production from 30 different soils was explained by the positive correlation to the ammonium or nitrate content. Less clear results were obtained, when only the temperature, but not the soil moisture, was kept constant (Baumgärtner and Conrad 1992b).

Another approach that is partially mechanistic, is to relate NO fluxes to a feature that is characteristic for a particular ecosystem and in some general way influences the NO flux in a mechanistic sense. A typical example is the net rate of nitrogen mineralization in soil which ultimately determines the availability of both ammonium and (after nitrification) nitrate in soil. Net mineralization thus should also determine in an integrative way the nitrogen transformation rates and thus NO flux. This approach was successfully used by Matson and Vitousek (1987) to obtain a significant correlation with N_2O emission rates across tropical forest ecosystems. However, NO flux and N_2O flux are often anticorrelated because they are controlled by soil moisture in different ways (Keller and Reiners 1994). Many ecosystems also obtain significant amounts of nitrogen through N_2 fixation or deposition (Melillo et al. 1989) so that net nitrogen mineralization rates are no longer limiting for nitrogen transformations. Recent laboratory measurements on different soils (n=30) did not give a significant correlation between rates of NO production (also of N_2O production) and net nitrogen mineralization (Bollmann and Conrad, unpublished). Despite these shortcomings, it is certainly useful to measure integrative features as potential indicators of NO flux, especially across well-defined landscapes, ecosystems and chronosequences (Keller and Reiners 1994).

Concluding remarks

The flux of NO between soil and atmosphere is very dynamic and varies over four orders of magnitude. The variability is due to large temporal and spatial variations in the soil conditions (e.g. texture, pH, temperature, moisture, organic carbon) that regulate microbial activities and due to the complexity of the controls (Fig.2).

1. The flux of NO into the atmosphere is regulated by the control of both production and consumption processes.
2. NO production is the result of nitrification, denitrification and other processes (including chemical reactions)
3. NO production is the result of the diversion of a small amount of NO (NO/product ratio) out of the general flow of nitrogen (N transformation rate)
4. Both N transformation rate and NO/product ratio are susceptible to regulation, and this regulation may be different for different microbial species
5. NO consumption is the result of denitrification and other processes (including chemical reactions)
6. NO consumption processes are susceptible to regulation, and this regulation may be different for different microbial species
7. The microbiological reactions that are potentially involved in NO metabolism include many different types of microorganisms with different metabolic pathways that are not all satisfactorily understood on a biochemical basis.
8. The NO-metabolizing microorganisms are phylogenetically highly diverse, even within one type of metabolic group (e.g. denitrifiers) and seem to exhibit different patterns of regulation.
9. There are presently no methods available which allow to determine the exact "species" (phylogenetically characterized) of NO-metabolizing microorganisms that are present in a particular soil.

References

Abeliovich A, Vonshak A (1992) Anaerobic metabolism of *Nitrosomonas europaea*. Arch Microbiol 158:267-270

Abou-Seada MNI, Ottow JCG (1988) Einfluß chemischer Bodeneigenschaften auf Ausmaß und Zusammensetzung der Denitrifikationsverluste drei verschiedener Bakterien. Z Pflanzenernaehr Bodenkd 151:109-115

Ahlers B, König W, Bock E (1990) Nitrite reductase activity in *Nitrobacter vulgaris*. FEMS Microbiol Lett 67:121-126

Alexander M, Marshall KC, Hirsch P (1960) Autotrophy and heterotrophy in nitrification. Proc Comm III, 7th Intern Congr Soil Sci, Madison (WI):586-591

Allison SM, Prosser JI (1993) Ammonia oxidation at low pH by attached populations of nitrifying bacteria. Soil Biol Biochem 25:935-941

Anderson IC, Levine JS (1986) Relative rates of nitric oxide and nitrous oxide production by nitrifiers, denitrifiers, and nitrate respirers. Appl Environ Microbiol 51:938-945

Anderson IC, Poth M, Homstead J, Burdige D (1993) A comparison of NO and N_2O production by the autotrophic nitrifier *Nitrosomonas europaea* and the heterotrophic nitrifier *Alcaligenes faecalis*. Appl Environ Microbiol 59:3525-3533

Aulakh MS, Doran JW, Mosier AR (1992) Soil denitrification - significance, measurement, and effects of management. Adv Soil Sci 18:1-57

Barraclough D, Puri G (1995) The use of ^{15}N pool dilution and enrichment to separate the heterotrophic and autotrophic pathways of nitrification. Soil Biol Biochem 27:17-22

Baumgärtner M (1991) Umsetzung von Stickoxiden (NO_x) in Böden, auf Gebäudeoberflächen und in Mikroorganismen. Konstanzer Dissertationen Nr. 327, Hartung-Gorre Verlag, Konstanz

Baumgärtner M, Conrad R (1992a) Role of nitrate and nitrite for production and consumption of nitric oxide during denitrification in soil. FEMS Microbiol Ecol 101:59-65

Baumgärtner M, Conrad R (1992b) Effects of soil variables and season on the production and consumption of nitric oxide in oxic soils. Biol Fertil Soils 14:166-174

Bedard C, Knowles R (1989) Physiology, biochemistry, and specific inhibitors of CH_4, NH_4^+, and CO oxidation by methanotrophs and nitrifiers. Microbiol Rev 53:68-84

Bell LC, Ferguson SJ (1991) Nitric and nitrous oxide reductases are active under aerobic conditions in cells of *Thiosphaera pantotropha*. Biochem J 273:423-427

Bender M, Conrad R (1994) Microbial oxidation of methane, ammonium and carbon monoxide, and turnover of nitrous oxide and nitric oxide in soils. Biogeochem 27:97-112

Betlach MR, Tiedje JM (1981) Kinetic explanation for accumulation of nitrite, nitric oxide, and nitrous oxide during bacterial denitrification. Appl Environ Microbiol 42:1074-1084

Blackmer AM, Cerrato ME (1986) Soil properties affecting formation of nitric oxide by chemical reactions of nitrite. Soil Sci Soc Am J 50:1215-1218

Blösl M, Conrad R (1992) Influence of an increased pH on the composition of the nitrate-reducing microbial populations in an anaerobically incubated acidic forest soil. Syst Appl Microbiol 15:624-627

Bock E, Koops HP, Harms H, Ahlers B (1991) The biochemistry of nitrifying organisms. In: Shively JM, Barton LL (eds) Variations in Autotrophic Life, Academic Press, London, pp 171-200

Bock E, Schmidt I, Stuven R, Zart D (1995) Nitrogen loss caused by denitrifying *Nitrosomonas* cells using ammonium or hydrogen as electron donors and nitrite as electron acceptor. Arch Microbiol 163:16-20

Bremner JM, Blackmer AM, Waring SA (1980) Formation of nitrous oxide and dinitrogen by chemical decomposition of hydroxylamine in soils. Soil Biol Biochem 12:263-269

Brons HJ, Hagen WR, Zehnder AJB (1991) Ferrous iron dependent nitric oxide production in nitrate reducing cultures of *Escherichia coli*. Arch Microbiol 155:341-347

Burth I, Ottow JCG (1983) Influence of pH on the production of N_2O and N_2 by different denitrifying bacteria and *Fusarium solani*. In: Hallberg R (ed) Environmental Biogeochemistry, vol 35, Ecol. Bull. (Stockholm), pp 207-215

Castignetti D, Hollocher TC (1981) Vigorous denitrification by a heterotrophic nitrifier of the genus *Alcaligenes*. Curr Microbiol 6:229-231

Castignetti D, Hollocher TC (1984) Heterotrophic nitrification among denitrifiers. Appl Environ Microbiol 47:620-623

Chalk PM, Smith CJ (1983) Chemodenitrification. Dev Plant Soil Sci 9:65-89

Cole JA (1988) Assimilatory and dissimilatory reduction of nitrate to ammonia. In: Cole JA, Ferguson S (eds) The Nitrogen and Sulphur Cycles, Cambridge University Press, Cambridge, pp 281-329

Conrad R (1990) Flux of NO_x between soil and atmosphere: Importance and soil microbial metabolism. In: Revsbech NP, Soerensen J (eds) Denitrification in Soil and Sediment, Plenum, New York, pp 105-128

Conrad R (1994) Compensation concentration as critical variable for regulating the flux of trace gases between soil and atmosphere. Biogeochem 27:155-170

Conrad R (1995) Soil microbial processes involved in production and consumption of atmospheric trace gases. Adv Microb Ecol, in press.

Coyne MS, Arunakumari A, Averill BA, Tiedje JM (1989) Immunological identification and distribution of dissimilatory heme cd_1 and nonheme copper nitrite reductases in denitrifying bacteria. Appl Environ Microbiol 55:2924-2931

Crutzen PJ (1979) The role of NO and NO_2 in the chemistry of the troposphere and stratosphere. Ann Rev Earth Planet Sci 7:443-472

Davidson EA (1991) Fluxes of nitrous oxide and nitric oxide from terrestrial ecosystems. In: Rogers JE, Whitman WB (eds) Microbial Production and Consumption of Greenhouse Gases: Methane, Nitrogen Oxides, and Halomethanes, American Society for Microbiology, Washington D.C., pp 219-235

Davidson EA (1992) Sources of nitric oxide and nitrous oxide following wetting of dry soil. Soil Sci Soc Am J 56:95-102

Davidson EA (1993) Soil water content and the ratio of nitrous oxide to nitric oxide emitted from soil. In: Oremland RS (ed) Biogeochemistry of Global Change, Chapman & Hall, New York, pp 369-386

Davidson EA, Hackler JL (1994) Soil heterogeneity can mask the effects of ammonium availability on nitrification. Soil Biol Biochem 26:1449-1453

Davidson EA, Hart SC, Shanks CA, Firestone MK (1991) Measuring gross nitrogen mineralization, immobilization, and nitrification by ^{15}N isotopic pool dilution in intact soil cores. J Soil Sci 42:335-349

Davidson EA, Matson PA, Vitousek PM, Riley R, Dunkin K, Garcia-Mendez G, Maass JM (1993) Processes regulating soil emissions of NO and N_2O in a seasonally dry tropical forest. Ecology 74:130-139

DeBoer W, Klein-Gunnewiek PJA, Troelstra SR, Laanbroek HJ (1989) Two types of chemolithotrophic nitrification in acid heathland humus. Plant and Soil 119:229-235

DeBoer W, Klein-Gunnewiek PJA, Veenhuis M, Bock E, Laanbroek HJ (1991) Nitrification at low pH by aggregated chemolithotrophic bacteria. Appl Environ Microbiol 57:3600-3604

Dendooven L, Anderson JM (1994) Dynamics of reduction enzymes involved in the denitrification process in pasture soil. Soil Biol Biochem 26:1501-1506

Duggin JA, Voigt GK, Bormann FH (1991) Autotrophic and heterotrophic nitrification in response to clear-cutting northern hardwood forest. Soil Biol Biochem 23:779-787

Firestone MK, Davidson EA (1989) Microbiological basis of NO and N_2O production and consumption in soil. In: Andreae MO, Schimel DS (eds) Exchange of Trace Gases between Terrestrial Ecosystems and the Atmosphere. Dahlem Konferenzen, Wiley, Chichester, pp 7-21

Firestone MK, Tiedje JM (1979) Temporal change in nitrous oxide and dinitrogen from denitrification following onset of anaerobiosis. Appl Environ Microbiol 38:673-679

Focht DD, Verstraete W (1977) Biochemical ecology of nitrification and denitrification. Adv Microb Ecol 1:135-214

Freitag A, Bock E (1990) Energy conservation in *Nitrobacter*. FEMS Microbiol Lett 66:157-162

Freitag A, Rudert M, Bock E (1987) Growth of *Nitrobacter* by dissimilatory nitrate reduction. FEMS Microbiol Lett 48:105-111

Galbally IE, Johansson C (1989) A model relating laboratory measurements of rates of nitric oxide production and field measurements of nitric oxide emission from soils. J Geophys Res 94:6473-6480

Galbally IE, Roy CR (1978) Loss of fixed nitrogen from soils by nitric oxide exhalation. Nature 275:734-735

Groeneweg J, Sellner B, Tappe W (1994) Ammonia oxidation in *Nitrosomonas* at NH_3 concentrations near K_m: effects of pH and temperature. Water Res 28:2561-2566

Groffman PM, Rice CW, Tiedje JM (1993) Denitrification in a tallgrass prairie landscape. Ecology 74:855-862

Hart SC, Nason GE, Myrold DD, Perry DA (1994) Dynamics of gross nitrogen transformations in an old-growth forest: the carbon connection. Ecology 75:880-891

Hayatsu M (1993) The lowest limit of pH for nitrification in tea soil and isolation of an acidophilic ammonia oxidizing bacterium. Soil Sci Plant Nutr 39:219-226

Head IM, Hiorns WD, Embley TM, McCarthy AJ, Saunders JR (1993) The phylogeny of autotrophic ammonia-oxidizing bacteria as determined by analysis of 16S ribosomal RNA gene sequences. J Gen Microbiol 139:1147-1153

Højberg O, Revsbech NP, Tiedje JM (1994) Denitrification in soil aggregates analyzed with microsensors for nitrous oxide and oxygen. Soil Sci Soc Am J 58:1691-1698

Hooper AB (1968) A nitrite-reducing enzyme from *Nitrosomonas europaea*. Preliminary characterization with hydroxylamine as electron donor. Biochim Biophys Acta 162:49-65

Hooper AB (1989) Biochemistry of the nitrifying lithoautotrophic bacteria. In: Schlegel HG, Bowien B (eds) Autotrophic Bacteria, Springer, Berlin, pp 239-265

Hutchinson GL, Brams EA (1992) NO versus N_2O emissions from NH_4^+-amended Bermuda grass pasture. J Geophys Res 97:9889-9896

Hutchinson GL, Guenzi WD, Livingston GP (1993) Soil water controls on aerobic soil emission of gaseous nitrogen oxides. Soil Biol Biochem 25:1-9

Hutchinson GL, Davidson EA (1993) Processes for production and consumption of gaseous nitrogen oxides in soil. In: Harper LA et al. (ed) Agricultural Ecosystem Effects on Trace Gases and Global Climate Change, American Society of Agronomy, Madison (WI), pp 79-93

Ignarro LJ, Fukuto JM, Griscavage JM, Rogers NE, Byrns RE (1993) Oxidation of nitric oxide in aqueous solution to nitrite but not nitrate - Comparison with enzymatically formed nitric oxide from L-arginine. Proc Natl Acad Sci USA 90:8103-8107

Ji XB, Hollocher TC (1988) Reduction of nitrite to nitric oxide by enteric bacteria. Biochem Biophys Res Comm 157:106-108

Johansson C (1989) Fluxes of NO_x above soil and vegetation. In: Andreae MO, Schimel DS (eds) Exchange of Trace Gases between Terrestrial Ecosystems and the Atmosphere. Dahlem Konferenzen, Wiley, Chichester, pp 229-246

Johansson C, Galbally IE (1984) Production of nitric oxide in loam under aerobic and anaerobic conditions. Appl Environ Microbiol 47:1284-1289

Kalkowski I, Conrad R (1991) Metabolism of nitric oxide in denitrifying *Pseudomonas aeruginosa* and nitrate-respiring *Bacillus cereus*. FEMS Microbiol Lett 82:107-112

Keller M, Reiners WA (1994) Soil atmosphere exchange of nitrous oxide, nitric oxide, and methane under secondary succession of pasture to forest in the atlantic lowlands of Costa Rica. Global Biogeochem Cycles 8:399-409

Kharitonov VG, Sundquist AR, Sharma VS (1994) Kinetics of nitric oxide autoxidation in aqueous solution. J Biol Chem 269:5881-5883

Killham K (1986) Heterotrophic nitrification. In: Prosser JI (ed) Nitrification, IRL-Press, Oxford, pp 117-126

Kobayashi M, Shoun H (1995) The copper-containing dissimilatory nitrite reductase involved in the denitrifying system of the fungus *Fusarium oxysporum*. J Biol Chem 270:4146-4151

Körner H (1993) Anaerobic expression of nitric oxide reductase from denitrifying *Pseudomonas stutzeri*. Arch Microbiol 159:410-416

Krämer M, Baumgärtner M, Bender M, Conrad R (1990) Consumption of NO by methanotrophic bacteria in pure culture and in soil. FEMS Microbiol Ecol 73:345-350

Krämer M, Conrad R (1991) Influence of oxygen on production and consumption of nitric oxide in soil. Biol Fertil Soils 11:38-42

Kreitinger JP, Klein TM, Novick NJ, Alexander M (1985) Nitrification and characteristics of nitrifying microorganisms in an acid forest soil. Soil Sci Soc Am J 49:1407-1410

Kuenen JG, Robertson LA (1994) Combined nitrification-denitrification processes. FEMS Microbiol Rev 15:109-117
Kuenen JG, Robertson LA (1988) Ecology of nitrification and denitrification. In: Cole JA, Ferguson S (eds) The Nitrogen and Sulphur Cycles, Cambridge University Press, Cambridge, pp 161-218
Lang E, Jagnow G (1986) Fungi of a forest soil nitrifying at low pH values. FEMS Microbiol Ecol 38:257-265
LinneVonBerg KHL, Bothe H (1992) The distribution of denitrifying bacteria in soils monitored by DNA-probing. FEMS Microbiol Ecol 86:331-340
Lipschultz F, Zafiriou OC, Wofsy SC, McElroy MB, Valois FW, Watson SW (1981) Production of NO and N_2O by soil nitrifying bacteria. Nature 294:641-643
Lipton SA, Choi YB, Pan ZH, Lei SZZ, Chen HSV, Sucher NJ, Loscalzo J, Singel DJ, Stamler JS (1993) A redox-based mechanism for the neuroprotective and neurodestructive effects of nitric oxide and related nitroso-compounds. Nature 364:626-632
Lloyd D (1993) Aerobic denitrification in soils and sediments: from fallacies to facts. TREE 8:352-356
Logan JA (1983) Nitrogen oxides in the troposphere: global and regional budgets. J Geophys Res 88:10785-10807
Malinski T, Taha Z (1992) Nitric oxide release from a single cell measured in situ by a porphyrinic-based microsensor. Nature 358:676-678
Martikainen PJ, DeBoer W (1993) Nitrous oxide production and nitrification in acidic soil from a Dutch coniferous forest. Soil Biol Biochem 25:343-347
Matson PA, Vitousek PM (1987) Cross-system comparisons of soil nitrogen transformations and nitrous oxide flux in tropical forest ecosystems. Global Biogeochem Cycles 1:163-170
McKenney DJ, Drury CF, Findlay WI, Mutus B, McDonnell T, Gajda C (1994) Kinetics of denitrification by *Pseudomonas fluorescens* - oxygen effects. Soil Biol Biochem 26:901-908
McKenney DJ, Lazar C, Findlay WJ (1990) Kinetics of the nitrite to nitric oxide reaction in peat. Soil Sci Soc Am J 54:106-112
Meixner FX (1994) Surface exchange of odd nitrogen oxides. Nova Acta Leopoldina 70:299-348
Melillo JM, Steudler PA, Aber JD, Bowden RD (1989) Atmospheric deposition and nutrient cycling. In: Andreae MO, Schimel DS (eds) Exchange of Trace Gases between Terrestrial Ecosystems and the Atmosphere. Dahlem Konferenzen, Wiley, Chichester, pp 263-280
Moncada S, Higgs EA (1991) Review - Endogenous nitric oxide - Physiology, pathology and clinical relevance. Eur J Clin Investig 21:361-374
Mulder A, VandeGraaf AA, Robertson LA, Kuenen JG (1995) Anaerobic ammonium oxidation discovered in a denitrifying fluidized bed reactor. FEMS Microbiol. Ecol. 16:177-184
Munch JC (1989) Organism specific denitrification in samples of an Udifluvent with different nitrate concentrations. Z Pflanzenernaehr Bodenkd 152:395-400
Nägele W, Conrad R (1990) Influence of soil pH on the nitrate-reducing microbial populations and their potential to reduce nitrate to NO and N_2O. FEMS Microbiol Ecol 74:49-58
Nielsen TH, Revsbech NP (1994) Diffusion chamber for nitrogen-15 determination of coupled nitrification-denitrification around soil manure interfaces. Soil Sci Soc Am J 58:795-800

Orso S, Gouy M, Navarro E, Normand P (1994) Molecular phylogenetic analysis of *Nitrobacter* spp. Int J Syst Bact 44:83-86

Padmaja S, Huie RE (1993) The reaction of nitric oxide with organic peroxyl radicals. Biochem Biophys Res Com 195:539-544

Papen H, Hellmann B, Papke H, Rennenberg H (1993) Emission of N-oxides from acid irrigated and limed soils of a coniferous forest in Bavaria. In: Oremland RS (ed) Biogeochemistry of Global Change, Chapman & Hall, New York, pp 245-260

Papen H, VonBerg R, Hinkel I, Thoene B, Rennenberg H (1989) Heterotrophic nitrification by *Alcaligenes faecalis*: NO_2^-, NO_3^-, N_2O, and NO production in exponentially growing cultures. Appl Environ Microbiol 55:2068-2072

Parkin TB (1993) Spatial variability of microbial processes in soil - a review. J Environ Qual 22:409-417

Parkin TB, Sexstone AJ, Tiedje JM (1985) Adaption of denitrifying populations to low soil pH. Appl Environ Microbiol 49:1053-1056

Paul JW, Beauchamp EG, Zhang X (1993) Nitrous and nitric oxide emissions during nitrification and denitrification from manure-amended soil in the laboratory. Can J Soil Sci 73:539-553

Pennock DJ, VanKessel C, Farrell RE, Sutherland RA (1992) Landscape-scale variations in denitrification. Soil Sci Soc Am J 56:770-776

Poth M (1986) Dinitrogen production from nitrite by a *Nitrosomonas* isolate. Appl Environ Microbiol 52:957-959

Remde A, Conrad R (1990) Production of nitric oxide in *Nitrosomonas europaea* by reduction of nitrite. Arch Microbiol 154:187-191

Remde A, Conrad R (1991a) Metabolism of nitric oxide in soil and denitrifying bacteria. FEMS Microbiol Ecol 85:81-93

Remde A, Conrad R (1991b) Production and consumption of nitric oxide by denitrifying bacteria under anaerobic and aerobic conditions. FEMS Microbiol Lett 80:329-332

Remde A, Conrad R (1991c) Role of nitrification and denitrification for NO metabolism in soil. Biogeochem 12:189-205

Remde A, Ludwig J, Meixner FX, Conrad R (1993) A study to explain the emission of nitric oxide from a marsh soil. J Atm Chem 17:249-275

Renault P, Sierra J, Stengel P (1994) Transfert d'oxygène et anoxie dans les sols agrégés: contribution à l'étude de la dénitrification. Agronomie 14:395-409

Risgaard-Petersen N, Rysgaard S, Nielsen LP, Revsbech NP (1994) Diurnal variation of denitrification and nitrification in sediments colonized by benthic microphytes. Limnol Oceanogr 39:573-579

Robertson GP (1989) Nitrification and denitrification in humid tropical ecosystems: Potential controls on nitrogen retention. In: Proctor J (ed) Mineral Nutrients in Tropical Forest and Savanna Ecosystems, Blackwell Scientific, Boston, pp 55-69

Robertson LA, Cornelisse R, Zeng R, Kuenen JG (1989) The effect of thiosulphate and other inhibitors of autotrophic nitrification on heterotrophic nitrifiers. Ant Leeuwenhoek 56:301-309

Rudolph J, Conrad R (1995) Flux between soil and atmosphere, vertical concentration profiles in soil, and turnover of nitric oxide: 2. Experiments with naturally layered soil cores. J Atm Chem, submitted.
Rudolph J, Rothfuss F, Conrad R (1995) Flux between soil and atmosphere, vertical concentration profiles in soil, and turnover of nitric oxide: 1. Measurements on a model soil core. J Atm Chem, submitted.
Rysgaard S, Risgaard-Petersen N, Nielsen LP, Revsbech NP (1993) Nitrification and denitrification in lake and estuarine sediments measured by the ^{15}N dilution technique and isotope pairing. Appl Environ Microbiol 59:2093-2098
Saad OALO, Conrad R (1993a) Adaptation to temperature of nitric oxide-producing nitrate- reducing bacterial populations in soil. Syst Appl Microbiol 16:120-125
Saad OALO, Conrad R (1993b) Temperature dependence of nitrification, denitrification, and turnover of nitric oxide in different soils. Biol Fertil Soils 15:21-27
Schäfer F, Conrad R (1993) Metabolism of nitric oxide by *Pseudomonas stutzeri* in culture and in soil. FEMS Microbiol Ecol 102:119-127
Schimel JP, Firestone MK, Killham KS (1984) Identification of heterotrophic nitrification in a Sierran forest soil. Appl Environ Microbiol 48:802-806
Schumacher W, Kroneck PMH, Pfennig N (1992) Comparative systematic study on "*Spirillum*" 5175, *Campylobacter* and *Wolinella* species - description of "*Spirillum*" 5175 as *Sulfurospirillum deleyianum* gen. nov., spec. nov. Arch Microbiol 158:287-293
Schumacher W, Kroneck PMH (1992) Anaerobic energy metabolism of the sulfur-reducing bacterium "*Spirillum*" 5175 during dissimilatory nitrate reduction to ammonia. Arch Microbiol 157:464-470
Schuster M, Conrad R (1992) Metabolism of nitric oxide and nitrous oxide during nitrification and denitrification in soil at different incubation conditions. FEMS Microbiol Ecol 101:133-143
Sexstone AJ, Revsbech NP, Parkin TB, Tiedje JM (1985) Direct measurement of oxygen profiles and denitrification rates in soil aggregates. Soil Sci Soc Am J 49:645-651
Shepherd MF, Barzetti S, Hastie DR (1991) The production of atmospheric NO_x and N_2O from a fertilized agricultural soil. Atmos Environ 25A:1961-1969
Shoun H, Kim DH, Uchiyama H, Sugiyama J (1992) Denitrification by fungi. FEMS Microbiol Lett 94:277-281
Shoun H, Tanimoto T (1991) Denitrification by the fungus *Fusarium oxysporum* and involvement of cytochrome P-450 in the respiratory nitrite reduction. J Biol Chem 266:11078-11082
Singh HB (1987) Reactive nitrogen in the troposphere. Chemistry and transport of NO_x and PAN. Environ Sci Technol 21:320-327
Skiba U, Smith KA, Fowler D (1993) Nitrification and denitrification as sources of nitric oxide and nitrous oxide in a sandy loam soil. Soil Biol Biochem 25:1527-1536
Slemr F, Seiler W (1991) Field study of environmental variables controlling the NO emissions from soil, and of the NO compensation points. J Geophys Res 96:13017-13031
Smith GB, Tiedje JM (1992) Isolation and characterization of a nitrite reductase gene and its use as a probe for denitrifying bacteria. Appl Environ Microbiol 58:376-384

Smith MS, Tiedje JM (1979) Phases of denitrification following oxygen depletion in soil. Soil Biol Biochem 11:261-267

Stamler JS, Singel DJ, Loscalzo J (1992) Biochemistry of nitric oxide and its redox-activated forms. Science 258:1898-1902

Stroo HF, Klein TM, Alexander M (1986) Heterotrophic nitrification in an acid forest soil and by an acid-tolerant fungus. Appl Environ Microbiol 52:1107-1111

Stüven R, Vollmer M, Bock E (1992) The impact of organic matter on nitric oxide formation by *Nitrosomonas europaea*. Arch Microbiol 158:439-443

Teraguchi S, Hollocher TC (1989) Purification and some characteristics of a cytochrome c-containing nitrous oxide reductase from *Wolinella succinogenes*. J Biol Chem 264:1972-1979

Teske A, Alm E, Regan JM, Toze S, Rittmann BE, Stahl DA (1994) Evolutionary relationships among ammonia- and nitrite-oxidizing bacteria. J Bacteriol 176:6623-6630

Tiedje JM (1988) Ecology of denitrification and dissimilatory nitrate reduction to ammonia. In: Zehnder AJB (ed) Biology of Anaerobic Microorganisms, Wiley, New York, pp 179-244

Tietema A, Wessel WW (1992) Gross nitrogen transformations in the organic layer of acid forest ecosystems subjected to increased atmospheric nitrogen input. Soil Biol Biochem 24:943-950

Tortoso AC, Hutchinson GL (1990) Contributions of autotrophic and heterotrophic nitrifiers to soil NO and N_2O emissions. Appl Environ Microbiol 56:1799-1805

VanCleemput O, Baert L (1976) Theoretical considerations on nitrite self-decomposition reactions in soils. Soil Sci Soc Am J 40:322-323

VandeGraaf AA, Mulder A, DeBruijn P, Jetten MSM, Robertson LA, Kuenen JG (1995) Anaerobic oxidation of ammonium is a biologically mediated process. Appl. Environ. Microbiol. 61:1246-1251.

VanNiel EWJ, Braber KJ, Robertson LA, Kuenen JG (1992) Heterotrophic nitrification and aerobic denitrification in *Alcaligenes faecalis* strain TUD. Ant Leeuwenhoek 62:231-237

VonSchulthess R, Kuhni M, Gujer W (1995) Release of nitric and nitrous oxides from denitrifying activated sludge. Water Res 29:215-226

Verhaegen K, VanCleemput O, Verstraete W (1988) Nitrification mediated nitrosation of organics and effects on soil microbial biomass. Biol Wastes 26:235-245

Verstraete W, Alexander M (1973) Heterotrophic nitrification in samples of natural ecosystems. Environ Sci Technol 7:39-42

Vosswinkel R, Neidt I, Bothe H (1991) The production and utilization of nitric oxide by a new, denitrifying strain of *Pseudomonas aeruginosa*. Arch Microbiol 156:62-69

Williams EJ, Hutchinson GL, Fehsenfeld FC (1992) NO_x and N_2O emissions from soil. Global Biogeochem Cycles 6:351-388

Wu QT, Knowles R, Niven DF (1994) O_2 regulation of denitrification in *Flexibacter canadensis*. Can J Microbiol 40:916-921

Ye RW, Averill BA, Tiedje JM (1994) Denitrification - production and consumption of nitric oxide. Appl Environ Microbiol 60:1053-1058

Yoshinari T (1980) N_2O reduction by *Vibrio succinogenes*. Appl Environ Microbiol 39:81-84

Zafiriou OC, Hanley QS, Snyder G (1989) Nitric oxide and nitrous oxide reduction and cycling during dissimilatory nitrite reduction by *Pseudomonas perfectomarina*. J Biol Chem 264:5694-5699

Zumft WG (1993) The biological role of nitric oxide in bacteria. Arch Microbiol 160:253-264

Microbiological Controls on Dimethylsulfide Emissions from Wetlands and the Ocean

Ronald P. Kiene[1]
Department of Marine Sciences
University of South Alabama
LSCB-25
Mobile AL 36688

Introduction

Interest in the biogeochemistry of dimethyl sulfide [CH_3SCH_3; DMS) has grown substantially in recent years because it is now recognized that DMS is the principal form of volatile sulfur which is emitted from natural ecosystems to the atmosphere (Andreae, 1990). The emissions of DMS from the ocean in particular have received a great deal of attention because the oceanic flux contributes nearly half of the global biogenic sulfur emission to the atmosphere (Andreae, 1990). Furthermore, it has been proposed that oceanic DMS emissions may play a role in modulating global climate through the generation of sulfate aerosols which scatter sunlight and contribute to cloud formation (Bates et al., 1987; Charlson et al., 1987; Malin et al., 1992). Wetland DMS emissions, on the other hand, have little effect on global fluxes but may be important with respect to regional atmospheric chemistry and to sulfur cycling within sulfur-poor freshwater ecosystems (Nriagu et al., 1987). Recent investigations by Hines and co-workers (de Mello and Hines, 1994; Hines, In Press) have found that *Sphagnum*-dominated wetlands, in particular, have areal DMS flux rates which are comparable to those observed from the oceans.

---

[1] also at Alabama Marine Environmental Sciences Consortium Dauphin Island Sea Lab, 101 Bienville Boulevard, Dauphin Island, AL 36528

NATO ASI Series, Vol. I 39
Microbiology of Atmospheric Trace Gases
Edited by J. Colin Murrell and Donovan P. Kelly

In both oceans and wetlands, the production of DMS appears to be the result of microbial transformations of organic sulfur compounds (Taylor, 1993; Taylor and Kiene, 1989). Recent studies have led to an improved understanding of how DMS is formed and destroyed in marine environments (Kiene, 1993; Taylor, 1993), but there is still much to be learned about what controls the net ecosystem production of this sulfur gas. In contrast to the advances made in marine DMS research, very little is known about how DMS is formed and catabolized under natural conditions in freshwater ecosystems.

I present below a brief overview of what is currently known about the microbiological and biogeochemical controls on DMS emissions from the marine environment and I highlight some important gaps in our knowledge of this subject. I will restrict most of my discussion to the marine water column rather than sediments or other habitats. The reader should see Visscher (this volume) for more detailed information on microbial transformations of DMS and its precursor DMSP in diverse marine habitats. As a contrast to work in the marine environment, I will also discuss a recent study which has investigated the microbial formation of DMS in anoxic, acidic peat bogs dominated by *Sphagnum* mosses (Kiene and Hines, *1995*). This work suggests that the major mechanism of DMS formation in peat is fundamentally different from that which predominates in marine systems. I will also describe some of the microbial and physico-chemical controls on DMS emissions from *Sphagnum*-dominated peatlands.

Microbial controls on DMS concentrations in seawater

It is the concentration of DMS in the surface layer of seawater which largely determines its flux across the sea surface (Andreae, 1990). Wind speed and temperature are also factors which affect the flux (Liss et al., 1993) but these act upon the dissolved pool to drive the flux. The amount of DMS dissolved in surface waters is a result of a dynamic balance between production and loss of DMS. Production of DMS is almost exclusively a biological process (see below for details), whereas losses can occur through biological oxidation as well as through physico-chemical processes such as sea-air exchange and photooxidation. The pathways leading to the formation and destruction of DMS in seawater are illustrated in Fig. 1.

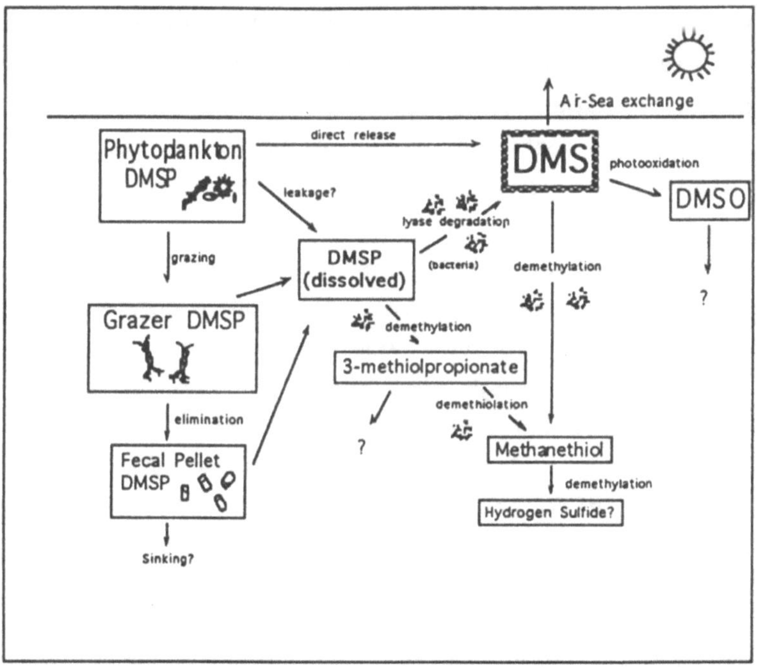

Figure 1. Pathways for DMSP and DMS in sea water. The diagram illustrates the role of phytoplankton DMSP as the major source of DMS and the multiple routes for liberation of dissolved DMSP and loss of DMS.

Degradation of the algal-derived organic sulfur compound DMSP [$(CH_3)_2S^+CH_2CH_2COOH$] appears to be the main source of DMS in sea water (see Fig. 1), although not all DMSP is degraded to DMS (Kiene and Service, 1991; Wolfe et al., 1994). The enzyme DMSP lyase is responsible for cleaving

DMSP into DMS and acrylic acid (Challenger and Simpson, 1948; de Souza and Yoch, 1995) and this enzyme has been found in bacteria and algae (de Souza and Yoch, 1995; Stefels and VanBoekel, 1993). It must be said that alternative precursors for DMS in aerobic seawater have not been investigated thoroughly, but it seems unlikely that other precursors will be significant due to the abundance of DMSP in phytoplankton and algae (Keller, 1989).

Most studies have found that DMS concentrations are not easily predicted from measures of algal biomass, primary productivity, DMSP distributions, or zooplankton biomass (Andreae, 1990). Only within nearly monospecific blooms of DMSP-producing algae (i.e. *E. huxleyi* and *Phaeocystis* spp.) do relationships between algal parameters and DMS become highly significant (Malin et al., 1993; Malin et al., 1992; Matrai and Keller, 1993). It is during such blooms of DMSP-producing algae that DMS concentrations may become elevated far above the typical background concentrations of <5 nM. The fluxes to the atmosphere from some of these blooms is large enough to have an impact on regional atmospheric sulfur chemistry (Malin et al., 1993). It is of interest then to understand what causes DMS to build up under certain circumstances while it remains low under others. This is an area where our understanding is still very poor.

The reasons for lack of strong correlations between algal biomass or DMSP and DMS under most conditions in the ocean include: 1) close coupling between DMS production and DMS loss; 2) losses of DMS due to biological and photochemical reactions; and 3) the fact that much of the DMSP undergoing degradation is converted to products other than DMS.

DMSP release and degradation, as well as the production of DMS are closely linked with food web activities (Dacey and Wakeham, 1986; Gabric et al., 1993; Kiene, 1993; Wolfe et al., 1994). Bacterial degradation of the dissolved DMSP pool (DMSP(d)) (operationally defined as that which passes a glass fiber or 0.2 µm filter) is thought to be a major pathway leading to DMS formation (Kiene, 1990; Kiene and Service, 1991; Turner et al., 1988). Lyase degradation of DMSP(d) is carried out by bacteria (Kiene, 1990; Ledyard and Dacey, 1994), and some phytoplankton (Stefels and VanBoekel, 1993). Under most conditions it is probably bacteria which are responsible for degrading the major fraction of DMSP(d) but further work is needed to document whether this is so.

The concentrations of DMSP(d) are generally in the low nM range in surface waters (Turner et al., 1988) and tend to be lower than the particulate (mostly algal) pool of DMSP (DMSP(p)). Despite typically low (1-4 nM) concentrations of DMSP(d), the kinetics of DMSP(d) degradation (and DMS formation) in natural water samples and in bacterial isolates show relatively low affinity for DMSP (apparent K_s > 100 nM) and a large maximum potential rate of degradation (V_{max}) (Kiene, 1992; Ledyard and Dacey, 1994). It is not clear at the present time whether this is the result of organisms naturally experiencing microzones of high [DMSP(d)] or whether the kinetics being measured reflect the combined effects of a cell-membrane transport enzyme system and an intracellular degradation system. Ledyard and Dacey (1994) presented data which suggested that DMSP must be transported into bacterial cells before it is degraded to DMS. Kiene and Gerard (in press) also provided evidence favoring an intracellular degradation of DMSP(d) in natural seawater samples. They found that a structural analog of DMSP, glycine betaine [$(CH_3)_3N^+CH_2COOH$], strongly inhibited DMSP(d) degradation in sea water from a variety of locations when added at 1000 fold higher concentrations than DMSP(d). These authors argued that glycine betaine may competitively inhibit the uptake of DMSP thereby preventing its degradation. Several other compounds with chemical structures similar to that of glycine betaine including proline, ß-alanine betaine, carnitine, and diethylsulfoniopropionate were also effective inhibitors, while structurally-unrelated compounds (i.e. glucose, glycine and acrylic acid) had no effects.

The ability to degrade DMSP is widespread among marine aerobic bacteria (Ledyard and Dacey, 1994; Visscher et al., 1992). At least 2 functional groups of bacteria are responsible for degrading DMSP in sea water:

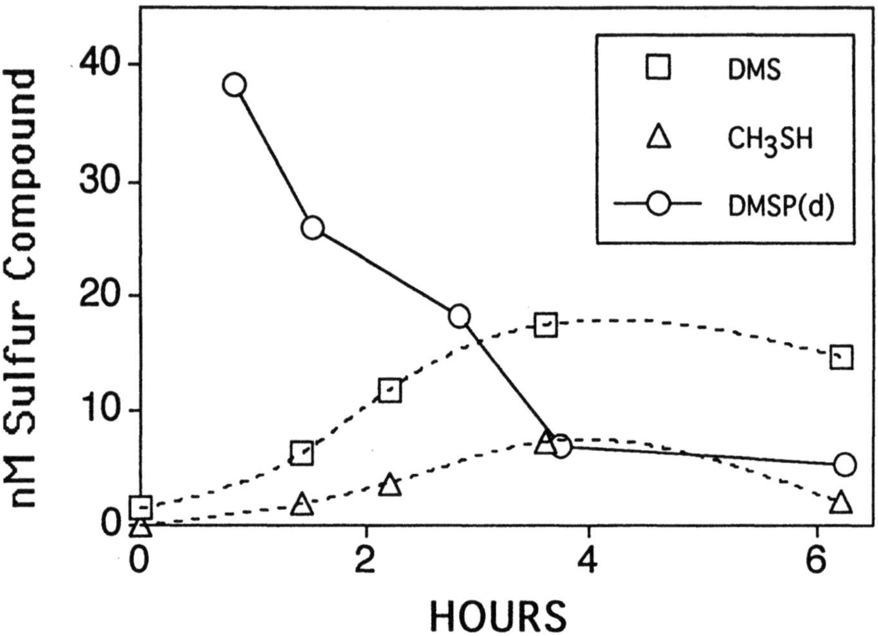

Figure 2. Time course of DMSP(d), DMS and MeSH in a seawater sample from Mobile Bay. DMSP(d) was added to an initial concentration of 38 nM at time = 1 h. The incubation temperature was 25 C. The results are typical of many similar experiments conducted with coastal and shelf waters.

those which cleave DMSP into DMS and acrylic acid and those which demethylate it to 3-methiolpropionate (Diaz et al.,

They argued that most of the DMSP which was degraded was converted to products other than DMS and they suggested that a demethylation pathway, which had recently been discovered in bacterial isolates (Taylor and Gilchrist, 1991), was the major route for DMSP metabolism (Fig. 2).

Recent results from my laboratory provide additional strong support for the idea that a major fraction of DMSP is being demethylated. We have observed significant production of methanethiol (MeSH) during the degradation of added DMSP(d)(Fig.2).

The relative yield of MeSH from DMSP in various coastal and shelf waters varied from 7 to 64% with a mean of 24% (Kiene, in preparation). This can be compared to the yield of DMS which ranged from 15 to 66% with a mean of 34%. Methanethiol arises from 3-methiolpropionate (See Fig. 1) which is the first demethylation product of DMSP (Taylor and Gilchrist, 1991). Methanethiol has been found to be very reactive (more so than DMS; Kiene, in preparation) and therefore the observed yields of MeSH significantly underestimate the fraction of DMSP undergoing demethylation. These data support the idea that, in most water samples, a major fraction (> 50%) of DMSP(d) is degraded without production of DMS (Fig. 2). The demethylation of DMSP may be one of the most important microbiological controls on the gross production of DMS in the oceans. There is evidently some very interesting microbial ecology involved in the control of DMS production. Factors which favor organisms which use either the lyase pathway or the demethylation pathway are currently unknown and are in need of investigation. Another dimension to this story is the fact that some of the bacteria which demethylate DMSP may also be responsible for metabolizing DMS (Visscher and Taylor, 1994).

Microbial metabolism of DMS is another major control on the concentration of this gas in seawater. In a study using surface waters from the Eastern Tropical Pacific Ocean, Kiene and Bates (1990) reported that biological degradation of DMS was a much more important sink for DMS in surface waters than was sea-air exchange. Lacking in that 1990 study was an assessment of the photochemical degradation which also removes DMS. Kieber et al. (Submitted) have now found that all three major DMS sinks (biological, photochemical and sea-air exchange) contributed significantly to the total DMS losses in the Equatorial Pacific. The relative importance of each sink varied considerably on a transect along 135°W longitude. At least for the biological sink, the reasons are not very clear as to why the rates varied. There is still much to be learned about what controls DMS metabolism in seawater.

To date there are still very few marine DMS-utilizing bacteria which have been isolated, especially those isolated from the water column. From the information available, there appears to be a diversity of DMS-degrading bacteria in the natural environment (Kelly and Smith, 1990; see also Visscher, this volume). Visscher and Taylor (1993) isolated several DMS-degrading bacteria with differing susceptibilities to inhibitors such as chloroform and methyl-butyl ether, further suggesting diversity in metabolic types.

Wolfe and Kiene (1993b) presented evidence from experiments with ^{14}C-DMS which suggested that DMS-carbon is both incorporated into cell material and respired to CO_2 in estuarine waters, with the percent respired ranging from 50 to 67%. DMS degradation in estuarine waters appeared to follow Michaelis-Menten-type kinetics with relatively low values of apparent K_s (< 20 nM) and V_{max} (< 18 nM DMS·d^{-1}). The metabolism of ^{14}C-DMS was inhibited by a variety of methylated and chlorinated organic compounds, including $CHCl_3$, dimethyldisulfide, and dimethylether (Wolfe and Kiene, 1993a). Judging by the few kinetic data available to date, it appears that there is considerably less capacity in seawater for DMS degradation than for DMSP degradation. Intense DMSP production (and subsequent degradation to DMS) will likely result in enhanced DMS accumulations because the DMS consumption will not keep pace with the production.

This is probably what happens in DMSP-producing algal blooms, although other factors may also be involved.

DMS dynamics in *Sphagnum* peat

Northern *Sphagnum*-wetlands are known to be significant sources of several trace gases including CH_4, CO_2 and DMS. Emissions of DMS from northern *Sphagnum*-wetlands can have an impact on regional atmospheric chemistry and the redistribution of sulfur among low sulfur freshwater ecosystems (Nriagu et al., 1987). In addition, loss of sulfur to the atmosphere prevents the sulfur from being sequestered in peat, which appears to be the fate of a large fraction of reduced sulfur in bogs (Brown, 1986; Wieder and Lang, 1988). The emission rates of DMS from the oceans (1-10 $\mu mol \cdot m^{-2}$-d^{-1}) and from *Sphagnum*-dominated wetlands (1-4 $\mu mol \cdot m^{-2}$-d^{-1}) are roughly similar when compared on an areal basis (Hines, In Press). Little has been published on the rates and pathways of DMS formation in natural freshwater systems and virtually no literature is available on DMS cycling in acidic bogs.

In collaboration with Dr. Mark Hines of the University of New Hampshire, I recently began an investigation into how DMS is formed in *Sphagnum* peat and what the factors are which control its release to the atmosphere. Below I summarize what we have found in these initial studies. Some of the data discussed below are also given in Kiene and Hines (*1995*).

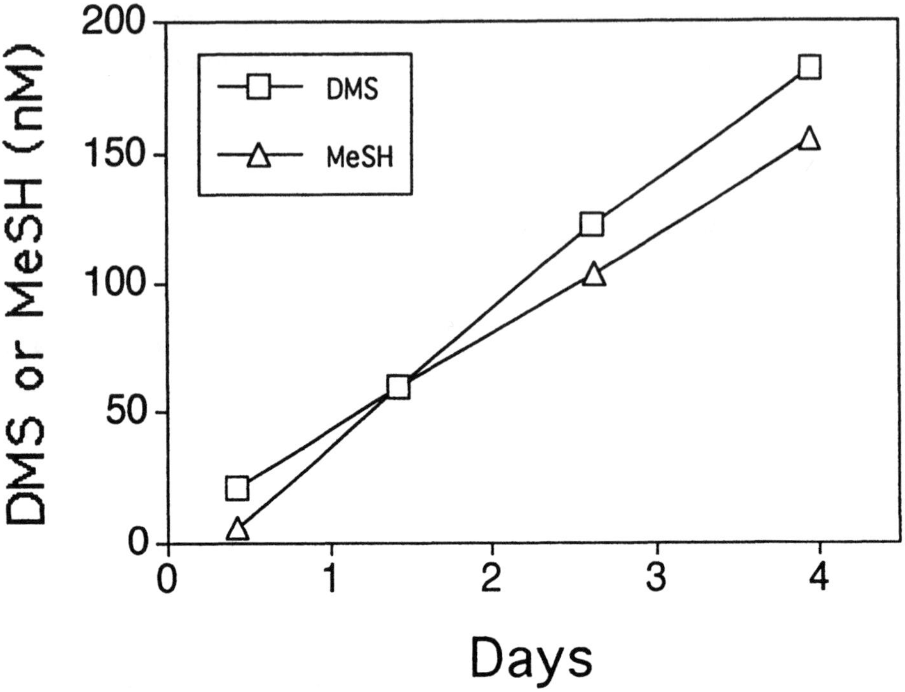

Figure 3. Time courses of DMS and MeSH concentrations in anoxic slurries of *Sphagnum* peat from Sallie's Fen. Both sulfur gases were produced at similar rates and the results are typical of at least 9 such slurry experiments conducted in summer 1993.

Experiments with anoxic slurries of *Sphagnum* peat have shown that both DMS and MeSH are produced at rather significant rates (Fig. 3). In most cases DMS production was linear and an average rate of 40.4 (range 22.0 to 68.6) nmol per liter of slurry·d^{-1} was observed in 9 batches of slurry. MeSH was produced at roughly similar rates over the typical 4-8 day incubations (Fig. 3). DMS and MeSH production in these acidic (pH 4.2-4.6) peats were biological as they were stopped completely by autoclaving and inhibited strongly by addition of antibiotics and 500 µM chloroform (Kiene and Hines, *1995*).

Several potential precursors of DMS in peat were identified; production of DMS was stimulated by 5-10 µM additions of S-methylmethionine, dimethylsulfoxide, or methoxyaromatic compounds (e.g. syringic acid). Similar concentrations of DMSP on the other hand did not stimulate DMS formation in these peats, in marked contrast to results obtained in seawater and marine sediments (Kiene and Service, 1991; Visscher and Van Gemerden, 1991; Visscher and Kiene, 1994). Since DMSP is not thought to be prevalent in freshwater systems it is not surprising that microorganisms do not efficiently degrade it in peat samples.

Despite the observed stimulation of DMS by dimethylated precursors and methoxyaromatics it was not clear which precursors or mechanisms of DMS formation are most important *in situ*. Based on experiments in which MeSH or methionine (a precursor of MeSH) were added we suspected that methylation of MeSH was one of the major pathways leading to DMS formation in peat. Thiol methylation activity has previously been observed in cultures of aerobic bacteria (Drotar et al., 1987) and in a variety of marine and freshwater anoxic sediments (Finster et al., 1990; Kiene and Visscher, 1987). Finster et al. (1990) suggested that methoxylated aromatic compounds such as syringic acid could be a source of methyl groups for the methylation of sulfide and subsequently MeSH. The results of Kiene and Hines (*1995*) support the potential role for syringic acid as a source of methyl groups in acidic peat but the importance of methoxyaromatic compounds relative to other methyl donors is currently unknown.

In order to investigate thiol methylation activity further, Kiene and Hines (*1995*) used a technique whereby thiol methylation activity was saturated with an analog of MeSH, namely ethanethiol (ESH). In this case the product of ESH methylation with endogenous methyl donors is ethylmethylsulfide (Fig. 4) which is easily distinguished from other sulfur gases in gas chromatography runs. Using this technique, Kiene and Hines showed that thiol methylation activity was saturated by 5 and 10 μM concentrations of ESH for several hours before abiological losses of ESH lowered the free concentration of the thiol.

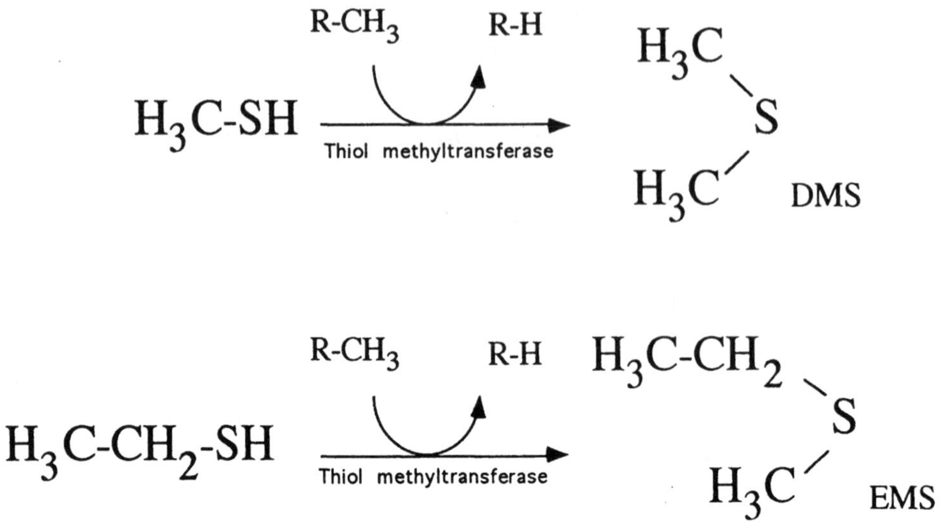

Figure 4. Scheme illustrating the methylation of thiols with natural methyl donors. Methanethiol is the dominant volatile thiol in natural anoxic environments and when methylated produces DMS. Ethanethiol is not abundant naturally, but can be added experimentally to study thiol methyltransferase activity. The product of ethanethiol methylation is ethylmethylsulfide, which is easily measured by conventional GC techniques.

The rates of thiol methylation activity reported in this study (173-200 nmol ethylmethylsulfide formed$\cdot l^{-1}\cdot d^{-1}$) greatly exceeded the average rate of DMS formation (~40 nmol DMS$\cdot l^{-1}\cdot d^{-1}$) observed in similar slurry incubations. It was concluded that methylation of natural MeSH with endogenous methyl donors was probably the main route of DMS formation in acidic *Sphagnum* peat.

A depth profile of thiol methylation activity in anoxically-incubated slurries and whole plant samples from Sallies Fen, New Hampshire is shown in Fig. 5. These data were taken in summer 1994 and were from samples which had been previously incubated for 4 days to obtain endogenous DMS production rates. ESH was then added to estimate thiol methylation rates. Maximum rates were about a factor of 10 higher than measured in fresh slurries the previous year (Kiene and Hines, *1995*) and were highest at 3 to 9 cm depth (Fig. 5). Consistent with the high methylation activity, these depths also showed the highest endogenous DMS and MeSH production

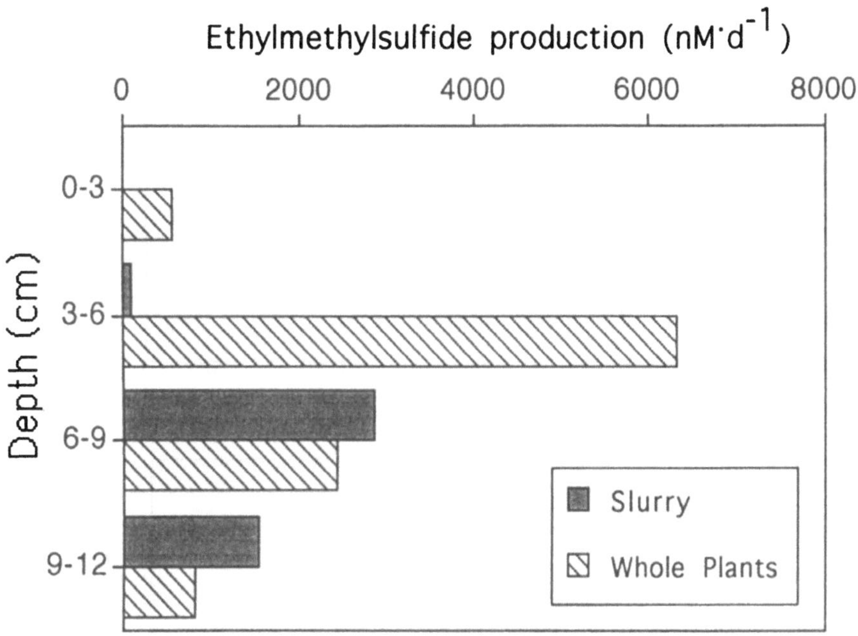

Figure 5. Depth profile of thiol (ESH) methylation activity in *Sphagnum* peat. In this case slurried samples were compared with unslurried whole *Sphagnum* plants in water. Methylation activity measurements were made on these samples after they had previously been incubated for 4 days to determine endogenous DMS and MeSH production rates. The prolonged incubation may have been responsible for the disparity in rates between slurries and whole plants at the 3-6 cm depth. This disparity was not seen in the DMS and MeSH production rates which were maximal at 6-9 cm.

rates (unpublished results). These observations are the first which point to the methylation of MeSH as the major mechanism for endogenous DMS production in any environment. The depth integrated rates of net DMS production observed in Sallie's Fen (6 $\mu mol \cdot m^{-2} \cdot d^{-1}$) are sufficient to explain the relatively high fluxes of DMS emitted to the atmosphere from this

methanogen inhibitor BES (Fig. 6). In the low pH peat, 2mM BES strongly inhibited DMS formation (Kiene and Hines, *1995*). It is possible but not likely that methanogens are directly

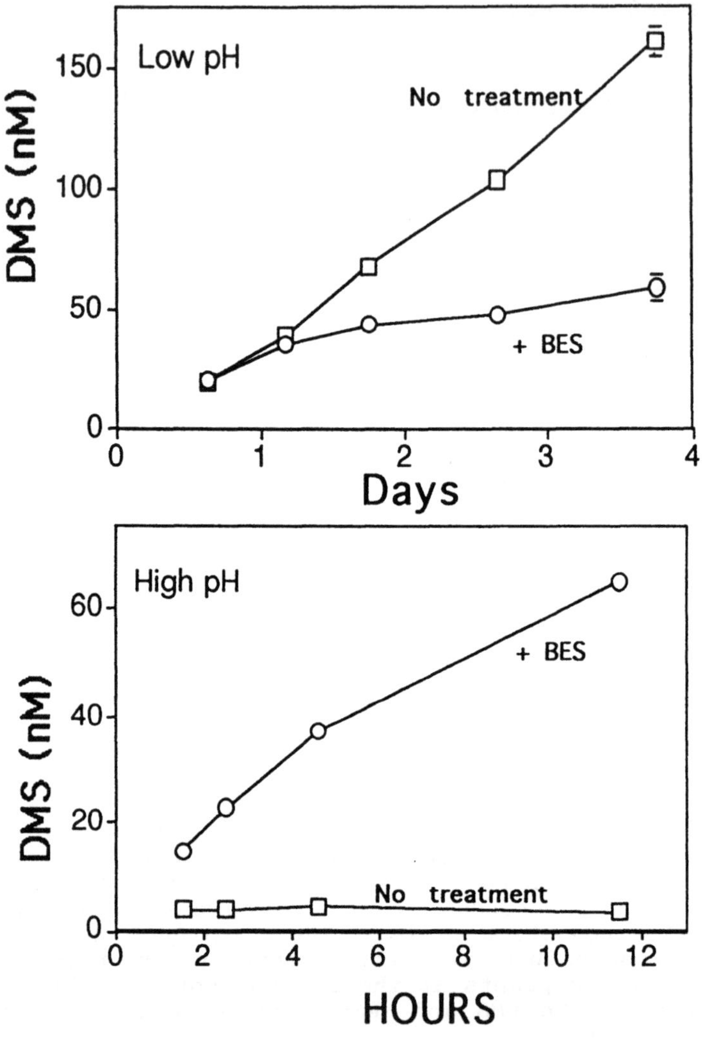

Figure 6. Contrasting effects of 2-bromoethanesulfonate (BES) on DMS accumulation in anoxic *Sphagnum* peat slurries from two sites in Sallie's Fen with different pH.

involved in DMS formation. A more likely explanation for these results is that BES caused acetate to accumulate which then inhibited acetogenic, thiol methylating bacteria. As stated above, acetate and other fermentable organic amendments were also inhibitory to DMS formation (Kiene and Hines, *1995*).

In the high pH peat, BES caused DMS to accumulate well above control levels (Fig. 6B) and it also inhibited losses of added DMS (not shown). It appears that methanogens might be consuming a major fraction of the DMS in high pH parts of the bog and that when methanogenesis is inhibited with BES, DMS accumulates. Results for MeSH were similar with BES causing significant stimulation of MESH accumulation. Methanogenic bacteria which use methylated sulfur compounds have been isolated (Ni and Boone, 1991; Oremland et al., 1989), but most are marine and none so far have come from acidic habitats. It should be noted that we characterize these sites by the differences in pH, but we do not know if pH is the controlling variable or whether other site differences are important.

The cycling of organic sulfur compounds in *Sphagnum* peat is presented schematically in Fig. 7. The important central role of MeSH is shown. MeSH may arise from conversion of methionine (Kiene and Hines, *1995*) or via the methylation of sulfide generated from sulfate reduction. The importance of the physico-chemical adsorption of MeSH to the solid phase of peat is indicated by the large arrow. This phenomenon effectively limits the formation of DMS, since it appears that endogenous methyl donors are abundant (see Fig. 5). Several dimethylated organic sulfur compounds could also be precursors of DMS in this system, but they are likely to be minor in comparison with MeSH methylation. Finally, the emission of DMS may be enhanced in low pH, oligotrophic sites due to slow or lacking consumption processes, which enable DMS to be transpired through the *Sphagnum* plants or to be emitted from standing bog water at relatively high rates.

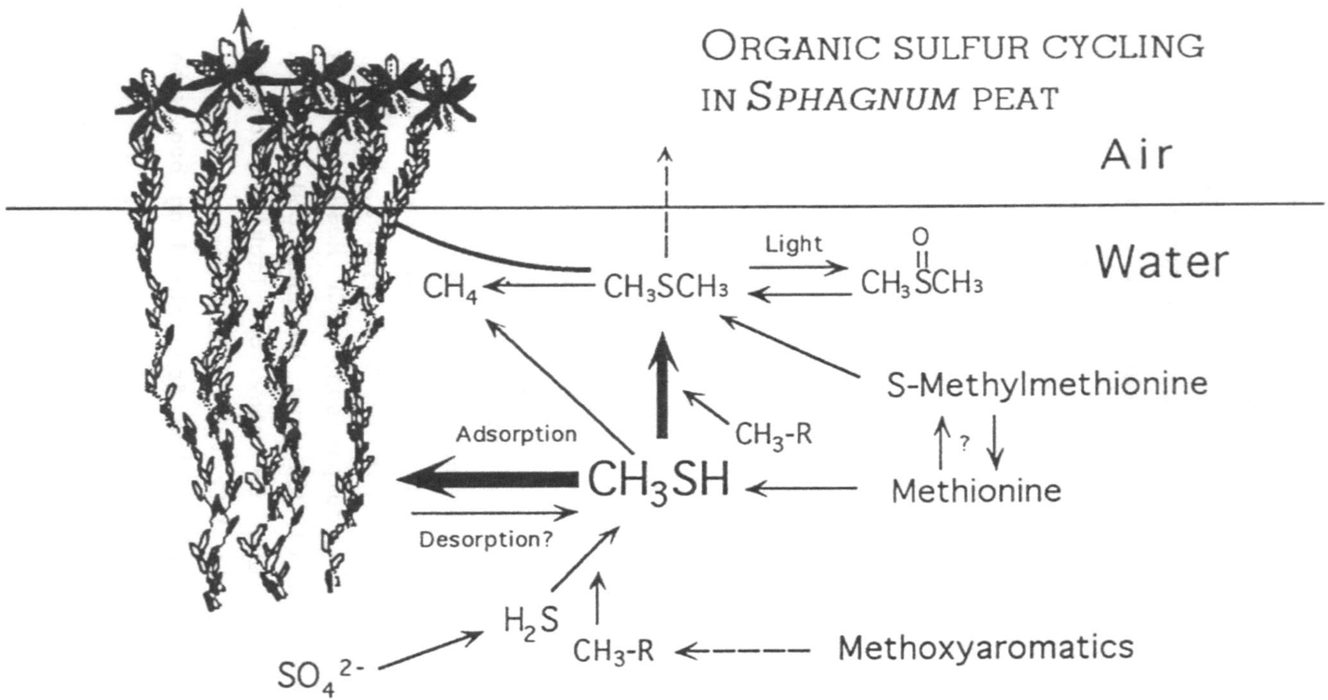

Figure 7. Scheme of organic sulfur compound cycling in anoxic *Sphagnum* peat based on results obtained in Sallies Fen. Potential exists for conversion of S-methylmethionine and DMSO to DMS, but it appears that methylation of endogenous MeSH is the major route leading to DMS formation. DMS is catabolised slowly in peat from low pH sites but may be consumed by methanogens in peat from higher pH areas. DMS emissions to the atmosphere likely occurs through the *Sphagnum* plants.

Summary

The microbiological and physico-chemical factors acting to control DMS emissions are quite different in the two systems discussed here. Methane thiol is an important product of organic sulfur metabolism in seawater and peatlands, but emissions of MeSH from either of these ecosystems are negligible due to chemical and biological reactivity of MeSH. Adsorption to surfaces and dissolved organic matter is probably the main sink for MeSH, although biological sinks also exist. Seawater DMS is derived almost exclusively from a dimethylated precursor, DMSP, whereas in the freshwater peat, the major route to DMS formation appears to be microbial methylation of MeSH. In both seawater and anoxic peat, microbial and physico-chemical reactions of the immediate precursors (DMSP and MeSH) greatly limit DMS formation. Further studies are needed to assess whether these findings can be generalized for all freshwater and marine habitats. Because of the importance of DMS fluxes to atmospheric chemistry and climate regulation, it will be of interest to study the microbiological controls on DMS emissions in further detail.

Acknowledgements

I am indebted to Mark Hines for a stimulating collaboration on DMS cycling in bogs. I also thank Ghislain Gerard for excellent technical assistance. Pieter Visscher and Barrie Taylor also deserve my gratitude for many years of stimulating collaboration. Financial support for this work was provided by NSF grants OCE-92-03728 and DEB-92-07978.

References

Andreae MO (1990) Ocean-atmosphere interactions in the global biogeochemical sulfur cycle. Mar Chem 30: 1-29

Bak F, Finster K, Rothfuss F (1992) Formation of dimethylsulfide and methanethiol from methoxylated aromatic compounds and inorganic sulfide by newly isolated anaerobic bacteria. Arch Microbiol 157: 529-534

Bates TS, Charlson RJ, Gammon RH (1987) Evidence for the climatic role of marine biogenic sulphur. Nature 329: 319-321

Brown KA (1986) Formation of organic sulphur in anaerobic peat. Soil Biol Biochem 18: 131-140

Challenger F, Simpson MI (1948) Studies on biological methylation, Part XII. A precursor of the dimethyl sulfide evolved by *Polysiphonia fastigiata*. Dimethyl-2-carboxyethyl sulfonium hydroxide and its salts. J Chem Soc 1948: 1591-1597

Charlson RJ, Lovelock JE, Andreae MO, Warren SG (1987) Oceanic phytoplankton, atmospheric sulfur, cloud albedo and climate. Nature 326: 655-661

Dacey JWH, Wakeham SG (1986) Oceanic dimethylsulfide: production during zooplankton grazing on phytoplankton. Science 233: 1314-1316

de Mello WZ, Hines ME Emissions of dimethylsulfide from northern *Sphagnum*-dominated wetlands: controls on production and flux. Glob. Biogeochem Cycles *Submitted*

de Mello WZ, Hines ME (1994) Application of static and dynamic enclosures for determining dimethylsulfide and carbonyl sulfide exchange in *Sphagnum* peatlands. J Geophys Res. 99:

de Souza MP, Yoch DC (1995) Purification and characterization of DMSP lyase from an *Alcaligenes*-like dimethylsulfide-producing marine isolate. Appl Environ Microbiol 61: 21-26

Diaz MR, Visscher PT, Taylor BF (1992) Metabolism of dimethylsulfoniopropionate and glycine betaine by a marine bacterium. FEMS Microbiol Lett 96: 61-66

Drotar A, Burton GA, Taverier JE, Fall R (1987) Widespread occurence of bacterial thiol methyltransferase and the biogenic emission of methylated sulfur gases. Appl Environ Microbiol 53: 1626-1631

Finster K, King GM, Bak F (1990) Formation of methylmercaptan and dimethylsulfide from methoxylated aromatic compounds in anoxic marine and fresh water sediments. FEMS Microbiol Ecol 74: 295-302

Gabric A, Murray N, Stone L, Kohl M (1993) Modelling the production of dimethylsulfide during a phytoplankton bloom. J Geophys Res 98: 22805-22816

Hines ME (In Press) Emissions of sulfur gases from wetlands. In: Adams D, Crill P, Seitzinger S (ed) Reduced gases in the hydrosphere. E Shweitzerbart'sche Verlagsbuchhandlungen, Stuttgart, Germany

Keller MD, Bellows WK, Guillard RRL (1989) Dimethyl sulfide production in marine phytoplankton. In: Saltzman E, Cooper WJ (ed) Biogenic sulfur in the environment. Am Chemical Soc, New York, p 167-182

Kelly DP, Smith NA (1990) Organic sulfur compounds in the environment: Bigeochemistry, microbiology, and ecological aspects. In: Marshall KC (ed) Advances in Microbial Ecology. Plenum Publishing Corp, New York, p 345-385

Kieber DJ, Jiao J, Kiene RP, Bates TS (Submitted) Impact of dimethylsulfide photochemistry on methyl sulfur cycling in the Equatorial Pacific Ocean. J Geophys Res Oceans

Kiene RP (1990) Dimethyl sulfide production from dimethylsulfoniopropionate in coastal seawater samples and bacterial cultures. Appl Environ Microbiol 56: 3292-3297

Kiene RP (1992) Dynamics of dimethyl sulfide and dimethylsulfoniopropionate in oceanic seawater samples. Mar Chem 37: 29-52

Kiene RP (1993) Microbial sources and sinks for methylated sulfur compounds in the marine environment. In: Kelly DP, Murrell JC (ed) Microbial Growth on C_1 Compounds.

Intercept Ltd, London, p 15-33

Kiene RP, Bates TS (1990) Removal of dimethylsulfide from seawater: Biological consumption versus sea- air exchange. EOS 71: 104

Kiene RP, Gerard G (*In press*) Evaluation of glycine betaine as an inhibitor of dissolved dimethylsulphoniopropionate degradation in marine waters. Mar Ecol Prog Ser

Kiene RP, Hines ME (*1995*) Microbial formation of dimethylsulfide in anoxic *Sphagnum* peat Appl Environ Microbiol. *In Press*

Kiene RP, Service SK (1991) Decomposition of dissolved DMSP and DMS in estuarine waters: dependence on temperature and substrate concentration. Mar Ecol Prog Ser 76: 1-11

Kiene RP, Visscher PT (1987) Production and fate of methylated sulfur compounds from methionine and dimethylsulfoniopropionate in salt marsh sediments. Appl Environ Microbiol 53: 2426-2434

Ledyard KM, Dacey JWH (1994) Dimethylsulfide production from dimethylsulfoniopropionate by a marine bacterium. Mar Ecol Prog Ser 110: 95-103

Liss P, Watson A, Liddicoat M, Malin G, Nightingale P, Turner S, Upstill-Goddard R (1993) Trace gases and air-sea exchanges. Phil Trans R Soc Lond A 343: 531-541

Malin G, Turner S, Liss P, Holligan P, Harbour D (1993) Dimethyl sulfide and dimethylsulphoniopropionate in the Northeast Atlantic during the summer coccolithophore bloom. Deep-Sea Res 40: 1487-1508

Malin G, Turner SM, Liss PS (1992) Sulfur: The plankton/climate connection. J Phycol 28: 590-597

Matrai PA, Keller MD (1993) Dimethylsulfide in a large-scale coccolithophore bloom in the Gulf of Maine. Cont Shelf Res 13: 831-843

Ni S, Boone DR (1991) Isolation and characterization of a dimethyl sulfide-degrading methanogen, *Methanolobus siciliae* HI350, from and oil well, characterization of *M. siciliae* T4/M^T, and emendation of *M. siciliae*. Intl J Syst Bacteriol. 41: 410-416

Nriagu JO, Holdway DA, Coker RD (1987) Biogenic sulfur and the acidity of rainfall in remote areas of Canada. Science 237: 1189-1192

Oremland RS, Kiene RP, Mathrani I, Whiticar MJ, Boone DR (1989) Description of an estuarine methylotrophic methanogen which grows on dimethyl sulfide. Appl Environ Microbiol 55: 994-1002

Stefels J, VanBoekel WHM (1993) Production of DMS from dissolved DMSP in axenic cultures of the marine phytoplankton species *Phaeocystis* sp Mar Ecol Prog Ser 97: 11-18

Taylor BF (1993) Bacterial transformations of organic sulfur compounds in marine environments. In: Oremland R (ed) Biogeochemistry of Global Change - Radiatively active trace gases. Chapman and Hall, New York, p 745-781

Taylor BF, Gilchrist DC (1991) New routes for aerobic biodegradation of dimethylsulfoniopropionate. Appl Environ Microbiol 57: 3581-3584

Taylor BF, Kiene RP (1989) Microbial metabolism of dimethyl sulfide. In: Saltzman ES, Cooper WJ (ed) Biogenic Sulfur in the Environment. American Chemical Society, Washington,DC, p. 202-221

Turner SM, Malin G, Liss PS (1988) The seasonal variation of dimethyl sulfide and dimethylsulfoniopropionate concentrations in nearshore waters. Limnol Oceanogr 33: 364-375

Visscher P, Van Gemerden H (1991) Production and consumption of dimethylsulfoniopropionate in marine microbial mats. Appl Environ Microbiol 57: 3237-3242

Visscher PT, Diaz MR, Taylor BF (1992) Enumeration of bacteria which cleave or demethylate dimethylsulfoniopropionate in the Caribbean Sea Mar Ecol Prog Ser 89: 293-296

Visscher PT, Kiene RP (1994) Demethylation and cleavage of dimethylsulfoniopropionate in marine intertidal sediments. FEMS Microbiol Ecol 14: 179-190

Visscher PT, Taylor BF (1993) A new mechanism for the aerobic catabolism of dimethyl sulfide. Appl Environ Microbiol 59: 3784-3789

Visscher PT, Taylor BF (1994) Demethylation of dimethylsulfoniopropionate to 3-mercaptopropionate by an aerobic marine bacterium. Appl Environ Microbiol 60: 4617-4619

Wieder RK, Lang GE (1988) Cycling of inorganic and organic sulfur in peat from Big Run Bog, West Virginia. Biogeochem 5: 221-242

Wolfe GV, Kiene RP (1993a) Effects of methylated, organic, and inorganic substrates on microbial consumption of dimethyl sulfide in estuarine waters. Appl Environ Microbiol 59: 2723-2726

Wolfe GV, Kiene RP (1993b) Radioisotope and chemical inhibitor measurements of dimethyl sulfide consumption rates and kinetics in estuarine waters. Mar Ecol Prog Ser 99: 261-269

Wolfe GV, Sherr EB, Sherr BF (1994) Release and consumption of DMSP from *Emiliania huxleyi* during grazing by *Oxyrrhis marina*. Mar Ecol Prog Ser 111: 111-119

Microbial Turn-over of Volatile Sulfur Compounds

Pieter T Visscher
Department of Marine Sciences
University of Connecticut
1084 Shennecossett Road
Groton CT 06340
USA

Ohne Schwefel kein Gedanke
after J. Moleschott (1822-1893)

Introduction

Biological production and consumption of sulfur gases include compounds such as H_2S, methylsulfides (CH_3SH, $(CH_3)_2S$, $(CH_3S)_2$), methanesulfonate (CH_3SO_3H), sulfoxides (SO_2), carbon disulfide (CS_2) and carbonyl sulfide (COS). Only H_2S, COS, SO_2 and CH_3SH are true gases under most environmental conditions, but other sulfur compounds, such as dimethylsulfide (DMS; $(CH_3)S$), dimethyldisulfide (DMDS; $(CH_3S)_2$), CS_2, and ethanethiol (ESH; C_2H_5SH) are liquids with vapor pressures which allow volatilization. Therefore, most of these sulfur compounds are gaseous and play a role in the global flux (estimated to be 210 Tg S y^{-1}; Table 1) from aquatic and terrestrial environments to the atmosphere. There, these gases engage in various (photo)chemical reactions, often mediated by radicals such as OH, NO_3 and IO (Plane 1989), and O_3 (Lee and Zhou 1994). Chemical reactions yield methanesulfonate (MSA), SO_2 and SO_4^{2-} (non-sea salt sulfate, nss). As in the sulfur cycle in the geosphere, atmospheric sulfur (S^{2-} to S^{4+}) is ultimately oxidized to sulfate (S^{6+}), and may return to the Earth's surface by wet and dry deposition. In addition to biogenic sources, anthropogenic sources also present a major source of sulfur gases (Table 1). Microbial sinks in terrestrial and aquatic

NATO ASI Series, Vol. I 39
Microbiology of Atmospheric Trace Gases
Edited by J. Colin Murrell and Donovan P. Kelly

Table 1. Estimated annual natural emission of volatile sulfur compounds‡

Compound	Flux (Tg/yr)	
H_2S	28.4	(6.4-70.6)
SO_2	13.1	(9.6-15)
COS	1.6	(0-3.5)
CS_2	1.8	(0.3-4.7)
DMS	46.5	(19.2-85)
SO_4^{2-}	179.2* £ 0.3	
Other (alkyl sulfides)	2.4	(1.2-3.6)
Total natural	94.1	(38-147)
Total anthropogenic	96	(78-121)
Total global	210	(116-268)

* sulfate emission in sea spray, most of which returns rapidly
‡ data from: Rodhe 1972; Bremner and Steele 1978; Whelpdale and Galloway 1979; Cullis and Hirschler 1980; Adams et al. 1981; Steudler and Peterson 1984; Rasmussen *et al.* 1984; Andreae 1990; Kelly and Smith 1990; Bates and Lamb 1992; Chin and Davis 1993

environments play an important role in determining the flux to the atmosphere. Due to logistics involved, direct atmospheric measurements of sulfur gases are scarce, but knowledge of biological and chemical reactions has increased significantly during the last decade. Several recent review papers have dealt with aspects of microbial metabolism of volatile sulfur compounds in detail (Kelly and Smith 1990, de Zwart and Kuenen 1992, Kiene 1993, Taylor 1993, Sorokin 1993). The scope of this contribution is to identify some of the processes leading to gaseous sulfur formation and to determine the potential role of bacteria in controlling the flux to the atmosphere by consumption of sulfur gases and their precursors.

Inorganic sulfur transformations - *Microbial physiology*

Sulfate, and other partially-reduced sulfur compounds (e.g. S^o, $S_2O_3^{2-}$, $S_nO_6^{2-}$), act as electron acceptors in anaerobic respiration. Dissimilatory reduction takes place via sulfate activation by ATP, catalyzed by ATP-sulfurylase. Adenosine phosphosulfate (APS) is then reduced to sulfite by APS reductase, and further to H_2S by sulfite reductase (possibly via polythionates and/or thiosulfate). Dissimilatory reduction of polysulfides and sulfur (Schauder and Müller 1993) as well as disproportionation of sulfite and thiosulfate (Bak and Pfennig 1987) also results in sulfide production. Assimilation of sulfate follows the same initial activation to APS. This is then either directly incorporated in sulfolipids, transformed by APS kinase to phosphoadenosine-phosphosulfate (PAPS) to yield sulf(on)ated polysaccharides, or directly reduced to an organic thionate ($R\text{-}S\text{-}SO_3^-$). The latter can be further reduced to organic sulfide ($R\text{-}S\text{-}S^-$), which can be incorporated in cysteine, and consecutively homocysteine, and S-adenosylmethionine or other organosulfur compounds such as dimethylsulfoniopropionate (DMSP).

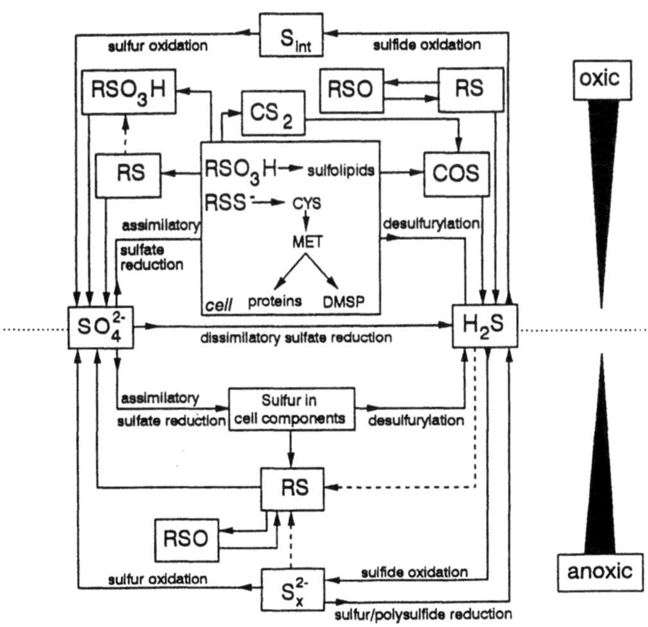

Figure 1. Biological pathways of sulfur. Solid lines indicate (micro)biological transformations; dashed lines display some of the important chemical reactions. R = organic moiety; S_{int} = reduced S intermediate; details of assimilation are similar under oxic and anoxic conditions. See text for details. (adapted from van Gemerden 1993)

Inorganic sulfide oxidation occurs under oxic and anoxic conditions (Figure 1). Chemotrophic and phototrophic bacteria presents the most important microbial sink (Jørgensen 1983). Oxidation of S^{2-} by chemotrophs is either coupled to oxygen or nitrate respiration to generate energy, or by phototrophs to produce reducing equivalents for photosynthesis. Sulfide is transformed to S^{o} or $S_x{}^{2-}$ by sulfide oxidase, which is further oxidized to sulfite (by sulfur oxidase) and sulfate (either by sulfite oxidase or APS reductase and ADP sulfurylase. Alternatively, sulfide can be oxidized to thiosulfate by flavocytochrome *c*. Thiosulfate is either transformed to trithionate (mediated by thiosulfate:acceptor oxidoreductase) or cleaved by rhodanese (or thiosulfate-sulfur transferase) to sulfur and sulfite before it is oxidized to sulfate.

Inorganic sulfur - *Microbial ecology and biogeochemistry*

Until two decades ago, H_2S was believed to be the main compound contributing to global sulfur flux from ocean and land to atmosphere (Rodhe 1972). Although it was concluded that this assumption was incorrect (Bremner and Steele 1978, Ingvorsen and Jørgensen 1982), reduced inorganic sulfur species are still key components in the production of sulfur-containing trace gases. In a Danish estuary, the H_2S emission exceeded the sulfate reduction rates and displayed a larger seasonal fluctuation (Ingvorsen and Jørgensen 1982). In the same study, the authors measured a H_2S flux of 0.4 µmol m^{-2} h^{-1} through an oxic waterbody of 0.7 m. Another study in Denmark (Jørgensen and Okholm Hansen 1985) reported a range for H_2S emission from 1 to 10 mmol S m^{-2} d^{-1}. Bioturbation may enhance the transport of sulfide from anoxic sediment layers to the atmosphere. However, less than 0.001% of the total sulfide produced by dissimilatory sulfate reduction presumably escaped to the atmosphere (Ingvorsen and Jørgensen 1982).

Sulfate reduction may occur under both anoxic and (sub)oxic conditions (Marschall et al 1993) in marine sediments (Figure 1; Fründ and Cohen 1992, Visscher et al. 1992A, Jørgensen 1983). Generally, 35% of organic carbon in marine sediments is mineralized by dissimilatory sulfate reduction, although values of > 70% have been reported (Thode Andersen and Jørgensen 1989). The sulfur content of living organisms is typically 0.3 to 3% of their dry weight, and assimilation of sulfur (see above) and consecutive decomposition of S-containing organic matter also yields hydrogen sulfide (Figure 1). However, sulfide formation from organic matter contributes little (approximately 3%; Jørgensen 1983) to the total S^{2-} production. Rapid microbial consumption, in addition to chemical oxidation or immobilization, causes little of the produced sulfide to escape to the atmosphere. However, H_2S (< 0.1 to 1.1 nM) was found in surface oceanic waters (Cutter and Krahforst 1988), possibly as a result of sulfate reduction in marine snow (Shanks and Reeder 1993). The chemical half-life of sulfide in seawater is 25 h (Zhang and Millero 1993) and, additionally, a high potential for microbial oxidation exists. As a result, little of this production is expected to escape to the atmosphere.

A sedimentary sink for sulfide is its reaction with metals, predominantly ferric oxyhydroxides, which forms FeS and FeS_2 (Berner 1967). Pyrite and FeS reoxidation is chemically and biologically mediated, but occurs at low rates (Luther et al. 1986, Luther 1987). One of the intermediates of pyrite oxidation is polysulfide, which may react with organic carbon (Vairavamurthy et al. 1992) to form organic polysulfides that may act as precursor of volatile sulfur compounds (Figure 1). Similarly, biological H_2S oxidation to sulfate produces intermediates such as polysulfides, which are often a major sulfur pool in marine sediments (Luther et al 1986, Visscher and van Gemerden 1993). Recently it was demonstrated that sulfide is oxidized to thiosulfate by thiobacilli, even at extremely low O_2 concentration (O_2/S^{2-} ratio = 0.43; van den Ende and van Gemerden 1994). Thiosulfate is rarely observed at concentrations higher than a few mM, probably because of rapid turnover by sulfate reducers, thiobacilli, phototrophs and heterotrophs (Visscher et al 1992a, Jørgensen 1994). Typically, high rates of sulfate reduction observed in marine sediments are in balance with biological and geochemical consumption.

For example, high rates of biological sulfide production and consumption in a marine sediments and microbial mat are tightly coupled (Roden and Tuttle 1993; see below). As a consequence, the flux of this volatile sulfur species to the atmosphere is expected to be low, or zero. Sulfate reduction rates in mats range from 1 to 10 mmol cm^{-3} d^{-1} (Jørgensen 1994, Fründ and Cohen 1992, Visscher et al. 1992a), or approximately 2 to 20 mmol m^{-2} d^{-1}. Specific reduction rates, based on counts of sulfate reducers (Ramsing et al. 1993, Visscher et al. 1992a) range from 1 to 50 10^{-15} mol $cell^{-1}$ d^{-1} (Jørgensen and Bak 1991, Visscher et al. 1992) similar to values obtained in laboratory cultures (Ingvorsen and Jørgensen 1984). Enumeration of sulfide oxidizers (Visscher and van Gemerden 1993) in combination with their oxidation rates in laboratory experiments (de Wit and van Gemerden 1990, van den Ende and van Gemerden 1993) displays a potential to oxidize > 100 mmol m^{-2} d^{-1}, even under O_2-limiting conditions as encountered *in situ* (Visscher and van Gemerden 1993). Sulfide consumption in similar ecosystems, measured with an ion-specific electrode, ranged from 0.8-20 mmol cm^{-3} d^{-1} in slurried sediments (C. Holliger et al. unpublished results). Although these calculations are merely estimates, they illustrate the potential to minimize the emission of H_2S, which might still occur when production and consumption are not in balance.

Organic sulfur transformations - *biogeochemical production*

Assimilatory sulfate reduction represents a dominant pathway in biological organosulfur production (Figure 1; see above). Volatile organic sulfides are produced during decomposition of S-containing organic matter, such as sulfolipids, amino acids, the osmolyte DMSP, and thiophenes (Kadota and Ishida 1972, Kelly and Smith 1990, Taylor 1993). Dimethyl sulfide (DMS) and methanethiol (CH_3SH) are the most commonly found products (Kiene 1993; see also chapter by Kiene in this volume). Volatile organic sulfides are also formed under anoxic conditions during demethoxylation of aromatic compounds such as syringate (Figure 2; Finster et al 1990). Similar transmethylation reactions are possibly very common in nature (Drotar et al. 1987a, Kiene 1993, Kiene and Capone 1988): DMS and CH_3SH production from sulfide was demonstrated in cultures (Drotar et al 1987b, Bak et al 1992). Methylbromide destruction in anoxic soils and sediment slurries may proceed via CH_3SH and DMS (Figure 2; Oremland et al 1994; see also chapter by Oremland in this volume). In addition to H_2S and S_x^{2-}, alkylsulfides, such as ESH and butanethiol, are strong nucleophiles which react with organic matter (Mopper and Taylor 1986, Vairavamurthy et al 1992) and also with halogenated organocompounds such as methylbromide and methyliodide (Barbash and Reinhard 1989). The products of nucleophilic substitution include volatile sulfur compounds. Finally, dimethylsulfoxide (DMSO) is a precursor of DMS in salt marsh sediments (Kiene and Capone 1988), where it acts as electron acceptor (Figure 2; Zinder and Brock 1978a). Hyphomicrobia reduce DMSO to DMS, which is further metabolized to SO_4^{2-} and CO_2 (Suylen et al 1986; see below).

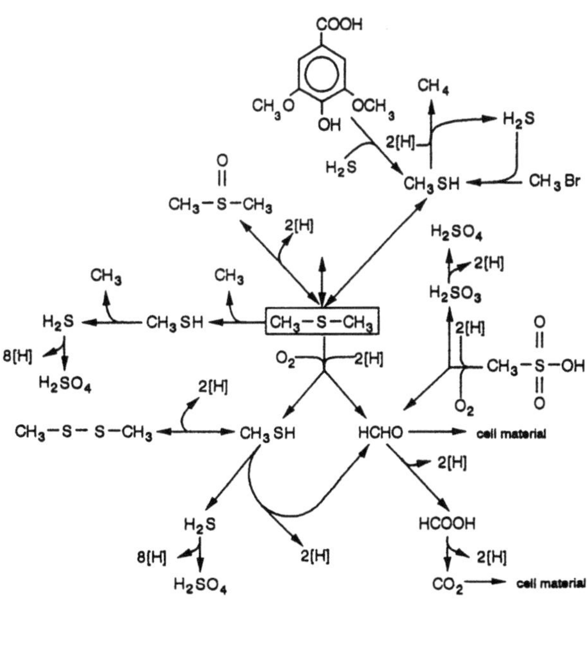

Figure 2. Microbial pathways of DMS, CH_3SH, DMSO, CH_3Br and MSA transformation (see text for explanation).

DMSP undergoes cleavage resulting in DMS formation (Figure 3). This process involves a DMSP-lyase, which is present in bacteria, phytoplankton and macroalgae. The other decomposition route involves demethylation/demethiolation via methiolpropionate, which yields CH_3SH as a volatile product (Taylor and Gilchrist 1991). This pathway of DMSP degradation was important in subtropical waters (Kiene 1992, Visscher et al. 1992b) Successive methylation of methanethiol produces DMS (see also chapter by Kiene in this volume).

Sulfur containing amino acids are precursors of DMS, DMDS and CH_3SH in anoxic soils (Francis et al 1975, Zinder and Brock 1978b, Kiene and Capone 1988). Enzymatic cleavage of methionine by methionine lyase yields CH_3SH (see Taylor 1993). Methylation of the CH_3SH produced from methionine in anoxic marine sediments may result in DMS formation (Kiene and Visscher 1987). In addition, 3-mercaptopropionate is formed during methionine and homocysteine metabolism in anoxic sediments (Kiene et al 1990).

COS and CS_2 are produced in a variety of environments, either by plants and other organisms, or (photo)chemically (see Kelly and Smith 1990, Taylor 1993). COS was formed when soils were amended with thiocyanate (SCN^-) (Bremner and Steele 1978). Emission of COS, presumably from biogenic origin, was reported for marine sediments (Jørgensen and Okholm-Hansen 1985). However, the presence of a biogenic (microbial) source in oceanic waters is disputed (Ferek and Andreae 1984). In a pure culture of *Thiobacillus thioparus*, COS is found in the headspace after SCN^- incubation (Katayama et al 1993). This organism employs a thiocyanate hydrolase to convert SCN^- to COS and ammonia.

Organic sulfur transformations - *biogeochemical consumption*

Consumption of volatile organic sulfur compounds by microorganisms includes oxidation of the sulfur moiety and demethylation reactions. The most common microbes involved inorganic sulfur transformations are sulfur bacteria (phototrophs and chemotrophs) and hyphomicrobia which engage in aerobic consumption of CH_3SH, DMS, DMDS, alkylsulfides

(ESH, propanethiol butanethiol), COS, CS_2 and MSA. Phototrophs, denitrifying thiobacilli, sulfate reducers and methanogens consume DMS, CH_3SH, ESH, CS_2 and DMDS under anoxic conditions. As is the case for H_2S (see above), a variety of bacteria that reach a significant population size in sediments have the potential to consume most, or all, of the DMS that is produced (Visscher and van Gemerden 1993). Although DMS emissions represent approximately 40% of the total flux in the USA (Bates and Lamb 1992) or 90% of the global sea/air exchange (Andreae 1990), most of its production in the water column is consumed by biogeochemical processes (Bates et al. 1994)

Metabolic transformations of DMS are outlined in Figure 2. The pathway of DMS consumption which employs a DMS monooxygenase is similar for hyphomicrobia (Suylen et al. 1986) and thiobacilli (Kanagawa and Kelly 1986, Smith and Kelly 1988b). The oxidation of DMS yields formaldehyde, which enters the serine pathway or is further oxidized via formaldehyde dehydrogenase to formate. The other product of initial DMS attack is CH_3SH, which is oxidized to formaldehyde and H_2S. Methanethiol oxidase from *Hyphomicrobium* and *T. thioparus* behaved very different, suggesting that the latter is better adapted to CH_3SH metabolism (Suylen et al. 1986, Gould and Kanagawa 1992). Microbial consumption of DMS under anoxic conditions, stimulated by nitrate, suggested another metabolic route. Inhibitor studies revealed an initial demethylation of DMS to CH_3SH, followed by a second demethylation reaction to H_2S and sulfate (Visscher and Taylor 1993). Similar demethylation reactions in which O_2 does not participate, may be common in marine methylotrophs (Visscher and Taylor 1994). Methylotrophic methanogens use the methyl group from DMS and CH_3SH to produce CH_4 (Oremland et al. 1989, Finster et al. 1992).

Slurry experiments in which molybdate was used, implied that sulfate reducers are important consumers of DMS under anoxic conditions (Kiene and Capone 1988). Recently, thermophilic sulfate reducers that use DMS and CH_3SH were described (Tanimoto and Bak 1994) but marine isolates are still lacking. Use of methyl groups from glycine betaine and

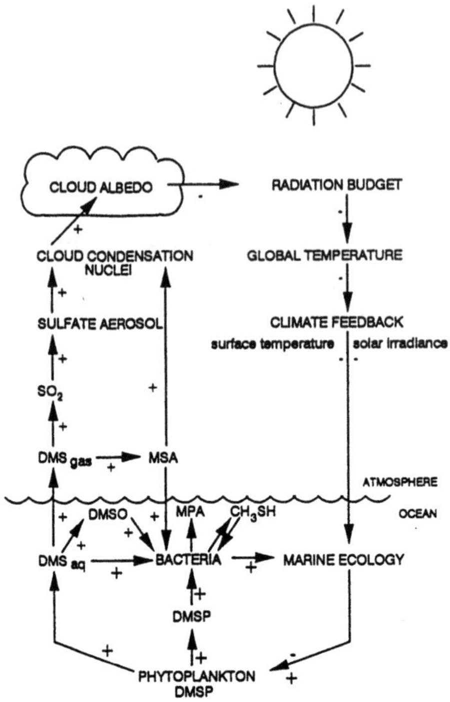

Figure 3. Production and consumption of DMS and implications for the global climate (adapted from Andreae 1990, Charlson 1987)

DMSP was demonstrated in cultures of marine sulfate reducers (Heijthuijsen and Hansen 1989, van der Maarel et al 1993). At elevated concentrations however, DMS may be a non-competitive substrate when it is used exclusively by methanogens (Kiene and Visscher 1987).

CS_2 is metabolized by *T. thioparus* Tk-m (Smith and Kelly 1988c), *Thiothrix ramosa* (Odinstova et al. 1993) and acts as the sole energy substrate in thiobacillus-like isolates obtained from soil and leaves around oak trees (Jordan et al. 1995). Of the latter, none were able to use thiocyanate nor methanethiol, and in contrast to the earlier isolates, were facultative heterotrophs. *T. thioparus* Tk-m also consumes COS (Smith and Kelly 1988c) and another *T. thioparus* THI 115 first produced COS in the headspace when growing on thiocyante, after which the culture incorporated this in cell material (Katayama et al 1993). Alternatively, COS undergoes hydrolysis which produces H_2S in oxygenated waters (Elliott et al. 1989). Although this is a relatively slow process, this may be another source of sulfide in the oceans.

Microbial degradation of MSA was found in methylotrophs isolated from soil (Baker et al. 1991). The absence of ribulose-bisphosphate carboxylase/oxygenase and presence of hydroxypyruvate reductase indicates the serine pathway for MSA oxidation. MSA monooxygenase yields formaldehyde, which is either oxidized to CO_2 or metabolized via the serine route. The sulfur moiety of MSA is oxidized to SO_4^{2-} via sulfite (Kelly et al. 1994).

Conclusion

Volatile sulfur compounds are emitted from land and oceans and are a major source of cloud condensation nuclei (CCN) (Charlson et al. 1987). Microbial consumption of precursors of nss has an important impact on the formation CCN and therefore, cloud formation and associated changes in global climate. In addition, bacterial destruction of atmospheric sulfur gases also functions as a potential controlling mechanism. The combination of chemical reactions, some of which occur at high rates, and microbial metabolism plays an important role in determining the concentration of trace gases in the atmosphere.

References

Adams DF, Farwell SO, Robinson E, Pack MR, Bamesberger WL (1981) Biogenic sulfur source strengths. Environ Sci Technol 15: 1493-1498

Andreae MO (1990) Ocean-atmosphere interactions in the global biogeochemical sulfur cycle. Mar Chem 30: 1-299

Bak F, Pfennig N (1987) Chemolithotrophic growth of *Desulfovibrio sulfodismutans* sp. nov. by disproportionation of inorganic sulfur compounds. Arch Microbiol 147: 184-189

Bak F, Finster K, Rothfuß P (1992) Formation of dimethylsulfide and methanethiol from methoxylated aromatic compounds and inorganic sulfide by newly isolated bacteria. Arch Microbiol 157: 529-534

Baker SC, Kelly DP, Murrell JC (1991) Microbial degradation of methanesulphonic acid: a missing link in the biogeochemical sulphur cycle. Nature 350: 627-628
Barbash JE, Reinhard M (1989) Reactivity of sulfur nucleophiles toward halogenated organic compounds in natural waters. In: Saltzman ES, Cooper WJ (eds) Biogenic sulfur in the environment. Am Chem Soc, Washington DC, pp 101-138
Bates TS, Lamb BK (1992) Natural sulfur emissions to the atmosphere of the continental United States. Global Biogeochem Cycles 6: 431-435
Bates TS, Kiene RP, Wolfe GV, Matrai PA, Chavez FP, Buck KR, Blomquist BW, Cuhel RL (1994) The cycling of sulfur in surface seawater of the northeast Pacific. J. Geophys. Res 99: 7835-7843
Berner RA (1967) Diagenesis of iron sulfide in recent marine sediments. In: Lauff GH (ed) Estuaries. Am Assoc Adv Sci, Washington DC, pp 268-272
Bremner JM , Steele CG (1978) Role of microorganisms in the atmospheric sulfur cycle. Adv Microb Ecol 2:155-201
Charlson RJ, Lovelock JE, Andreae MO, Warren SG (1987) Oceanic phytoplankton, atmospheric sulphur, cloud albedo and climate. Nature 326: 655-661
Chin M, Davis DD (1993) Global sources and sinks of OCS and CS_2 and their distributions. Global Biogeochem Cycles 7: 321-337
Cullis CF, Hirschler MM (1980) Atmospheric sulfur: natural and man-made sources. Atmos Environ 14:1263-1278
Cutter B and Krahforst D (1988) Sulfide in surface waters of the Western Atlantic Ocean. Geophys Res Lett 15: 1393-1396
de Wit R, van Gemerden H (1990) Growth of the phototrophic purple sulfur bacterium *Thiocapsa roseopersicina* under oxic/anoxic regimens in the dark. FEMS Microb Ecol 73:69-76
de Zwart JMM, Kuenen JG (1992) C_1-cycle of sulfur compounds. Biodegrad 3: 37-59
Drotar A, Burton Jr GA, Tavernier JE, Fall R (1987a) Widespread occurrence of bacterial thiol methyltransferases and biogenic emission of methylated sulfur gases. Appl Environ Microbiol 53: 1626-1631
Drotar A, Fall R, Mishalanie, Tavernier JE (1987b) Enzymatic methylation of sulfide, selenide, and organic thiols by *Tetrahymena thermophila*. Appl. Environ Microbiol 53: 2111-2118
Elliott S, Lu E, Sherwood-Rowland F (1989) Hydrogen sulfide in oxic seawater. In: Saltzman ES, Cooper WJ (eds) Biogenic sulfur in the environment. Am Chem Soc, Washington DC, pp 314-326
Ferek RJ, Andreae MO (1984) Photochemical production of carbonylsulphide in marine surface waters. Nature 307: 148-150
Finster K, King GM, Bak F (1990) Formation of methylmercaptan and dimethylsulfide from methoxylated aromatic compounds in anoxic marine and freshwater sediments. FEMS Microbiol Ecol 74: 295-302

Finster K, Tanimoto Y, Bak F (1992) Fermentation of methanethiol and dimethylsulfide by a newly isolated methanogenic bacterium. Arch Microbiol 157: 425-430

Francis AJ, Duxbury JM, Alexander M (1975) Formation of volatile organic products in soils under anaerobiosis. II Metabolism of amino acids. Soil Biol Biochem 7: 51-56

Fründ, C, Cohen Y (1992) Diurnal cycles of sulfate reduction under oxic conditions in cyanobacterial mats. Appl Environ Microbiol 58: 70-77

Gould WD, Kanagawa T (1992) Purification and properties of methyl mercaptan oxidase from *Thiobacillus thioparus* Tk-m. J Gen Microbiol 138: 217-221

Heijthuijsen JHFG, Hansen TA (1989) Anaerobic degradation of betaine by marine *Desulfobacterium* strains. Arch Microbiol 152: 393-396

Ingvorsen K, Jørgensen BB (1982) Seasonal variation in H_2S emission to the atmosphere from intertidal sediments in Denmark. Atmosph Environ 16: 855-865

Ingvorsen K, Jørgensen BB (1984) Kinetics of sulfate uptake by freshwater and marine species of *Desulfovibrio*. Arch Microbiol 139:61-66

Jordan SL, Kraczkiewicz-Dowjat AJ, Kelly DP, Wood AP (1995) Novel eubacteria able to grow on carbon disulfide. Arch Microbiol 163: 131-167

Jørgensen BB (1983) The microbial sulfur cycle. In: Krumbein WE (ed) Microbial geochemistry. Blackwell Scientific Publications, Oxford, UK, pp 91-124

Jørgensen BB (1994) Sulfate reduction and thiosulfate transformations in a cyanobacterial mat during a diel oxygen cycle. FEMS Microbiol Ecol 13: 303-312

Jørgensen BB, Okholm-Hansen B (1985) Emissions of biogenic sulfur gases from a Danish estuary. Atmosph Environ 19: 1737-1749

Jørgensen BB and Bak F (1991) Pathways and microbiology of thiosulfate transformations and sulfate reduction in a marine sediment (Kattegat, Denmark). Appl Env Microbiol 57:847-856

Kadota H, Ishida Y (1972) Production of volatile sulfur compounds by microorganisms. Ann Rev Microbiol 26:127-138

Kanagawa T, Kelly DP (1986) Breakdown of dimethyl sulphide by mixed cultures and by *Thiobacillus thioparus*. FEMS Microbiol Lett 34: 13-19

Katayama Y, Kanagawa T, Kuraishi H (1993) Emission of carbonyla sulfide by *Thiobacillus thioparus* grown with thiocyanate in pure and mixed cultures. FEMS Micro Letts 114:223-228

Kelly DP, Smith NA (1990) Organic sulfur compounds in the environment. Biogeochemistry, microbiology, and ecological aspects. Adv Microb Ecol 11:345-385

Kelly DP, Baker SC, Trickett J, Davey M, Murrell JC (1994) Methanesulphonate utilization by a novel methylotrophic bacterium involves in an unusual monooxygenase. Microbiol 140: 1419-1426

Kiene RP (1992) Dynamics of dimethyl sulfide and dimethylsulfoniopropionate in oceanic water samples. Mar Chem 37: 29-52

Kiene RP (1993) Microbial sources and sinks for methylated sulfur compounds in the marine environment. In: Kelly DP, Murrell, JC (eds) Microbial Growth on C_1 compounds, 7. Intercept, London, pp 15-33
Kiene RP, Visscher PT (1987) Production and fate of methylated sulfur compounds from methionine and dimethylsulfoniopropionate in salt marsh sediments. Appl Environ Microbiol 53: 2426-2434
Kiene RP, Capone DG (1988) Microbial transformations of methylated sulfur compounds in anoxic salt marsh sediments. Microbial Ecol 15: 275-291
Kiene RP, Malloy KD, Taylor BF (1990) Sulfur-containing amino acids as precursors of thiols in anoxic coastal sediments. Appl Environ Microbiol 56: 156-161
Lee Y-N, Zhou X (1994) Aqueous kinetics of ozone and dimethylsulfide and its atmospheric implications. J Geophys Res 99: 3597-3605
Luther GW III (1987) Pyrite oxidation and reduction: molecular orbit theory considerations. Geochim Cosmochim Acta 51: 3193-3199
Luther GW III, Church TM, Scudlark JR, Cosman M (1986) Inorganic and organic sulfur cycling in salt-marsh porewaters. Science 232: 746-749
Marschall C, Frenzel P, Cypionka H (1993) Influence of oxygen on sulfate reduction and growth of sulfate-reducing bacteria. Arch Microbiol 159: 168-173
Mopper K, Taylor BF (1986) Biogeochemical cycling of sulfur. In: Sohn M (ed) Organic marine geochemistry. Am Chem Soc, Washington DC, pp 324-339
Odintsova EV, Wood AP, Kelly DP (1993) Chemolithoautotrophic growth of *Thiothrix ramosa*. Arch Microbiol 160: 152-157
Oremland RS, Kiene RP, Mathrani I, Whiticar MJ, Boone DR (1989) Description of an estuarine methylotrophic methanogen which grows on dimethyl sulfide. Appl Environ Microbiol 55: 994-1002
Oremland RS, Miller LG, Strohmaier FE (1994) Degradation of methylbromide in anaerobic sediments. Environ Sci Technol 28:514-520
Plane JMC (1989) Gas-phase atmospheric oxidation of biogenic sulfur compounds. In: Saltzman ES and Cooper WJ (eds) Biogenic sulfur in the environment. Am Chem Soc, Washington DC, pp 404-423
Ramsing NB, Kühl M, Jørgensen (1993) Distribution of sulfate-reducing bacteria, O_2, H_2S in photosynthetic biofilms determined by oligonucleotide probes and microelectrodes. Appl Environ Microbiol 59: 3840-3849
Rasmussen RA, Khalil MAK, Crawford AJ, Fraser PJ (1982) Natural and anthropogenic trace gases in the Southern Hemisphere. Geophys Res Lett 9:704-707
Roden EE, Tuttle JH (1993) Inorganic sulfur turnover in oligohaline estuarine sediments. Biogeochem 22:81-105
Rodhe H (1972) A study of the sulfur budget for the atmosphere over northern Europe. Tellus 24: 128-138
Schauder R, Müller E (1993) Polysulfide as a possible substrate for sulfur-reducing bacteria. Arch Microbiol 160: 377-382

Shanks AL Reeder ML (1993) Reducing microzones and sulfide production in marine snow. Mar Ecol Prog Ser 96:43-47

Smith NA, Kelly DP (1988a) Isolation and physiological characterization of autotrophic sulfur bacteria oxidizing dimethyl disulphide as sole source of energy. J Gen Microbiol 134: 1407-1417

Smith NA, Kelly DP (1988b) Mechanisms of oxidation of dimethyl disulfide by *Thiobacillus thioparus* strain E6. J Gen Microbiol 134: 3031-3039

Smith NA, Kelly DP (188c) Oxidation of carbon disulfide as the sole source of energy for the autotrophic growth of *Thiobacillus thioparus* strain Tk-m. J Gen Microbiol 134: 3041-3048

Sorokin DYu (1993) Biological oxidation of sulfur atom in C_1-sulfur and other organosulfur compounds. Microbiol 62:575-581

Steudler PA, Peterson BJ (1984) Contribution of gaseous sulphur from salt marshes to the global sulfur cycle. Nature (London) 311:455-457

Suylen GMH, Stefess GC, Kuenen JG (1986) Chemolithotrophic potential of a *Hyphomicrobium* species capable of growth on methylated sulfur compounds. Arch Microbiol 146: 192-198

Tanimoto Y, Bak F (1994) Anaerobic degradation of methylmercaptan and dimethylsulfide by newly isolated thermophilic sulfate-reducing bacteria. Appl. Environ Microbiol 60: 2450-2455

Taylor BF (1993) Bacterial transformations of organic sulfur compounds in marine environments. In: Oremland, RS (ed) Biogeochemistry of global change. Chapman & Hall, New York, pp 745-781

Taylor BF, Gilchrist DC (1991) New routes for the aerobic biodegradation of dimethylsulfoniopropionate. Appl Environ Microbiol 57: 3581-3584

Thode Andersen S, Jorgensen BB (1989) Sulfate reduction and formation of 35S-labeled FeS, FeS2 and So in coastal marine sedimetns. Limnol Oceanogr 34:793-806

van den Ende FP, van Gemerden H (1993) Sulfide oxidation under oxygen limitation by a *Thiobacillus thioparus* isolated from a marine microbial mat. FEMS Microbiol Ecol 13: 69-78

van der Maarel MJEC, Quist P, Dijkhuizen L, Hansen TA (1993) Anaerobic degradation of dimethylsulfoniopropionate to 3-S-methylmercaptopropionate by a marine *Desulfobacterium* strain. Arch Microbiol 160: 411-412

van Gemerden H (1993) Microbial mats: a joint venture. Mar Geol 113: 3-25

Vairavamurthy A, Mopper K, Taylor BF (1992) Occurrence of particle bound polysulfides and significance of their reaction with organic matter in marine sediments. J Geophys Res Lett 19: 2043-2046

Visscher PT, Prins RA, van Gemerden H (1992a) Rates of sulfate reduction and thiosulfate consumption in a marine microbial mat. FEMS Microbiol Ecol 86: 283-294

Visscher PT, Diaz MR, Taylor BF (1992b) Enumeration of bacteria which cleave or demethylate

dimethylsulfoniopropionate in the Caribbean Sea. Mar Ecol Prog Ser 89:293-296
Visscher PT, van Gemerden H (1993) Sulfur cycling in laminated marine microbial ecosystems. In: Oremland RS (ed) Biogeochemistry of Global Change. Chapman & Hall, New York, pp 672-690
Visscher PT, Taylor BF (1993) A new mechanism for the aerobic catabolism of dimethyl sulfide, Appl Environ Microbiol 59: 3784-3789
Visscher PT, Taylor BF (1994) Demethylation of dimethylsulfoniopropionate to 3-mercaptopropionate by an aerobic marine bacterium. Appl Environ Microbiol 60: 4617-4619
Whelpdale DM, Galloway JN (1979) An atmospheric sulfur budget for Eastern North America. Atmos Environ 14:409-417
Zhang J-Z, Millero FJ (1993) The products from the oxidation of H_2S in seawater. Geochim Cosmochim Acta 57: 1705-1718
Zinder SH, Brock TD (1978a) Methane, carbon dioxide, and hydrogen sulfide production from the terminal methiol group of methionine by anaerobic lake sediments. Appl Environ Microbiol 35: 344-352
Zinder SH, Brock TD (1978b) Dimethyl sulfoxide reduction by microorganisms. J Gen Microbiol 105:335-342

Bacterial Metabolism of Methanesulfonic Acid

J Colin Murrell, Tim Higgins and Donovan P Kelly[1]
Department of Biological Sciences and [1]Institute of Education
University of Warwick
Coventry CV4 7AL
England

Introduction

The principal organic sulfur compound entering the atmosphere from aquatic and terrestrial environments is dimethyl sulfide (DMS). The majority of this is released into the oceans from the cleavage of dimethyl sulfoniopropionate, an osmoregulator found in phytoplankton such as *Emiliana* and *Phaeocystis* and the marsh grass *Spartina* (see chapters by Kiene and Visscher, this volume). Global estimates of DMS release vary, although it has been estimated that between 40-70 Tg is lost to the atmosphere annually representing a significant fraction of the global biogeochemical sulfur cycle. DMS in the atmosphere is photochemically oxidized to methanesulfonic acid (MSA) and sulfur dioxide. DMS is also oxidized to MSA by atmospheric NO_3 radicals in the dark (see Kelly *et al.* 1993), resulting in an annual global rate of MSA formation of 20-50 Tg MSA. MSA is known to be important in cloud condensation nuclei (i.e. it may contribute to cloud albedo) and therefore it may play an important role in global temperature control (Charlson *et al.* 1987). MSA is a very stable strong acid which is not further oxidized in the atmosphere but falls on the Earth's surface in rain and snow, thus re-entering the terrestrial and marine biosphere. Evidence for the continual deposition of MSA over thousands of years has been obtained from examination of Antarctic ice cores, although there is apparently no accumulation of MSA in the environment (Saigne and Legrand 1987). The inference then is that MSA is being mineralized in the environment to recycle the sulfur originating from DMS release in the oceans.

NATO ASI Series, Vol. I 39
Microbiology of Atmospheric Trace Gases
Edited by J. Colin Murrell and Donovan P. Kelly

Until recently, very little was known about the biodegradation of MSA. Its use as a sulfur source by algae and bacteria has been noted (summarized by Kelly *et al.* 1994) however the subsequent fate of the carbon released during this metabolism has not been investigated. We and our colleagues have isolated in our laboratories, bacteria from terrestrial, freshwater and marine environments that are capable of using MSA as a methylotrophic carbon source for growth. We briefly describe here the properties of these organisms, which appear to be ubiquitous in the environment and provide a missing link in the biogeochemical cycling of organic sulfur arising from DMS emissions from the oceans.

Isolation of MSA-degrading bacteria

Chemostat and batch culture enrichments using MSA as sole carbon and energy source, together with a methylotrophic co-substrate such as methylamine, methane or methanol yielded a number of MSA-degrading bacteria when various soils, freshwater, and marine samples were used as inocula (Baker *et al.* 1991; Thompson *et al.* 1995; Murrell and Affleck, unpublished). These organisms all appear to be restricted facultative methylotrophs which will grow on methanesulfonate as their carbon and energy source with doubling times of about 8-10 hours. The most well-characterized of these organisms is methylotroph Strain M2, a gram negative aerobic rod isolated from garden soil (Baker *et al.* 1991). It grew on MSA with a doubling time of 7-8 hr in batch culture and growth yields of 7-14 g dry wt. mol^{-1}).

It also grew aerobically on the one-carbon (C_1) compounds methylamine, formate, formaldehyde and methanol but not on the C_1 substrates methane, methane phosphonate, dimethyl sulfide dimethyl disulfide, nor the autotrophic energy source thiosulfate. It grew poorly on some organic acids and sugars (Baker *et al.* 1991; Baker 1992). Marine MSA utilizers have also been isolated from Plymouth Sound and the Tamar estuary (England). Two strains, TR3 and PSCH4, were examined in detail. They are both gram negative rods which grow aerobically on MSA as sole carbon and energy source. Like

Strain M2 they are restricted facultative methylotrophs but they have an obligate requirement for NaCl (0.5-3.5% w/v). They are somewhat more difficult to work with than terrestrial MSA-utilizers such as Strain M2 because they grow very poorly on agar, agarose or Phytogel plates. In liquid batch culture they grow with doubling times of 10-15 hr (Thompson *et al.* 1995).

Physiology and biochemistry of MSA utilization

Growth of Strain M2 on MSA was not chemolithotrophic or autotrophic and ribulose bisphosphate carboxylase could not be detected by enzyme assay, Western blotting or DNA probing techniques after growth on C_1 compounds. There was no hexulose phosphate synthase activity in cell-free extracts from Strain M2 grown on any C_1 substrates. High activities of hydroxypyruvate reductase were, however, found in cell extracts of strain M2 grown on all C_1 substrates, indicating that the Serine pathway was the primary route for formaldehyde assimilation. Formate dehydrogenase and methanol dehydrogenase activities were also found in cell-free extracts from Strain M2 grown on all C_1 compounds.

MSA oxidation by Strain M2 was an inducible trait. MSA-grown cells oxidized MSA, amino-MSA, ethanesulfonate and propanesulfonate but not higher alkylsulfonates, suggesting that a specific MSA-inducible oxygenase might be responsible for growth on MSA. Methanol, formaldehyde and formate were all oxidized by Strain M2 cells grown on MSA, methanol, formaldehyde, formate or methylamine but not after growth on pyruvate. Methylamine was only oxidized by cells previously grown on this substrate, suggesting that the methylamine oxidation pathway differs from that of MSA. Oxygen: substrate stoichiometries for C_1 substrates utilized by Strain M2 were 1.96, 1.43, 0.97 and 0.48 for MSA, methanol, formaldehyde and formate respectively, which is consistent with the complete oxidation of these substrates to carbon dioxide. One possibility for the mechanism of oxidation of MSA by Strain M2 was the hydrolysis of MSA to methanol and sulfurous acid:

$$CH_3SO_3H + H_2O = CH_3OH + H_2SO_3$$

with the methanol produced being further oxidized to formaldehyde by methanol dehydrogenase (MDH). This mechanism however, has been discounted since cyclopropanol, a specific inhibitor of pyrroloquinoline quinone linked MDH, allows the complete oxidation of MSA by MSA-grown cells of Strain M2, but inhibits methanol oxidation in the same cells (Kelly *et al.* 1994). Also several mutants of Strain M2 that are unable to grow on methanol have been isolated which have been shown to possess an inactive MDH but still grow on MSA (Higgins *et al.* 1995).

The presence of an NADH-dependent, MSA-specific monooxygenase (MSAMO; see below) in cell-free extracts of MSA-grown Strain M2 suggests that MSA degradation is initiated by a monooxygenase cleavage of MSA to yield formaldehyde and sulfite:

$$CH_3SO_3H + O_2 + NADH + H^+ = HCHO + NAD^+ + H_2SO_3 + H_2O$$

The sulfite produced is subsequently oxidized to sulfate while the further oxidation of formaldehyde to formate and then to CO_2 and water would yield (i) reducing power and energy for growth, (ii) NADH for the initial oxidation of MSA and (iii) some of the formaldehyde being assimilated into cell carbon via the serine pathway. The mechanism we propose for the metabolism of MSA by Strain M2 is outlined in Fig. 1. Characterization of the marine isolates TR3 and PSCH4 has shown that the pathway of MSA metabolism in these bacteria is likely to be very similar to that found in Strain M2 (Thompson *et al.* 1995).

Methanesulfonic acid monooxygenase (MSAMO)

Cell-free extracts of MSA-grown Strain M2 contain a cytoplasmic multi-component, NADH-specific monooxygenase enzyme system which we have called methanesulfonic acid monooxygenase (MSAMO). This is specifically induced during growth on MSA and has a narrow substrate specificity,

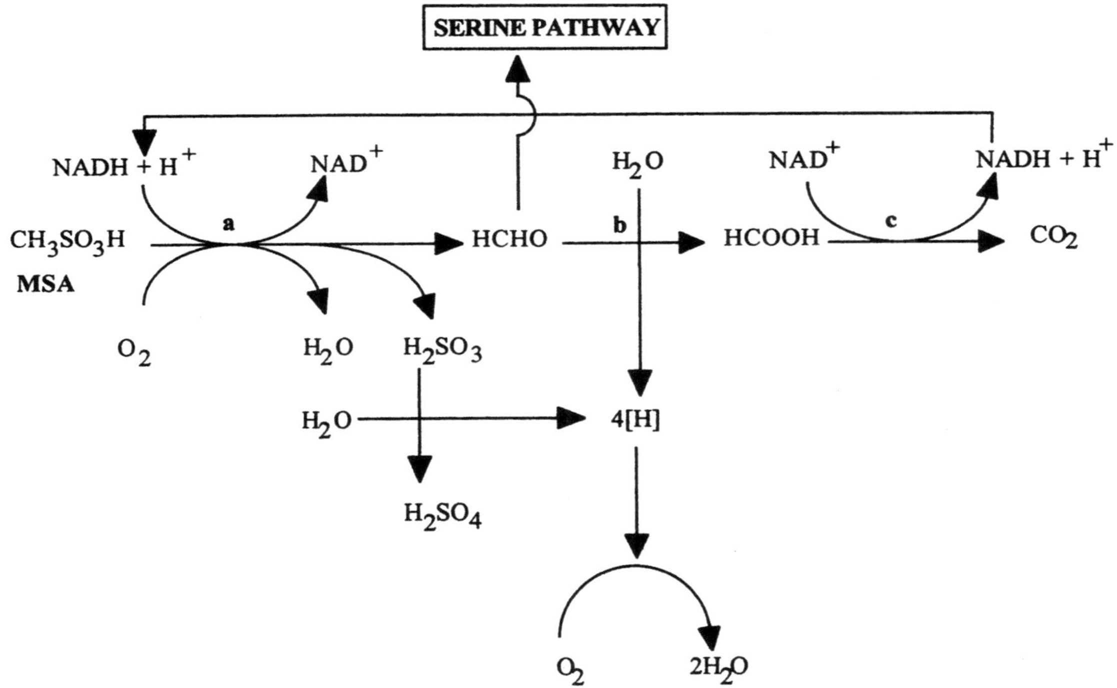

Figure 1. Proposed pathway for the oxidation of MSA by Strain M2. Key reactions are : (a) an NADH-dependent MSAMO yields formaldehyde and sulfite (which is subsequently oxidised to sulfate); (b) formaldehyde is subsequently assimilated via the serine pathway or is oxidised to formate by formaldehyde dehydrogenase; (c) formate is further oxidised to CO_2 by formate dehydrogenase, thus regenerating NADH.

oxidizing MSA and ethane- or propane-sulfonic acids but not butane- or pentane-sulfonic acids. The enzyme catalyses the following reaction:

$$CH_3SO_3H + O_2 + NADH + H^+ = HCHO + NAD^+ + H_2SO_3 + H_2O$$

Sulfite released in this reaction can be detected using Ellman's reagent (Johnston *et al.* 1975). This provides a quick and convenient alternative assay to the spectrophotometric assay based on MSA-stimulated consumption of NADH (Kelly *et al.* 1994) for initial detection of MSAMO during subsequent purification steps. MSAMO is a relatively stable enzyme in cell-free extracts. Activity is inhibited by metal chelators such as bathophenanthroline, bathocuproine, dipyridyl and EDTA, suggesting that the enzyme has associated metal ions, which is also a characteristic of other oxygenase systems (Mason and Cammack 1992). Enzyme activity was also inhibited by the respiratory chain inhibitors cyanide, azide and carbon monoxide, which indicated that electron transfer reactions were essential for C-S bond cleavage by MSAMO (Higgins *et al.* 1995).

The MSAMO has been resolved into three fractions, components A, B and C by Q-Sepharose ion exchange chromatography. Restoration of enzyme activity *in vitro* required all three fractions which when added together restored MSA-dependent NADH oxidation (as measured spectrophotometrically) and MSA-dependent sulfite production (determined using Ellman's reagent). Addition of FAD and Fe^{2+} to the reaction mixture resulted in a five-fold increase in enzyme activity, suggesting that both FAD and Fe^{2+} are required for activity and are being lost from the enzyme complex during the purification procedure. There was a high endogenous rate of NADH oxidation in the spectrophotometric assay when component B was absent, suggesting that component A or component C was the NADH-binding protein. Further purifications of component A have led to the identification of a 36-38 kDa monomeric protein which can reduce a number of electron acceptors such as cytochrome c, nitroblue tetrazolium and ferricyanide in the presence of NADH, suggesting that this is the putative reductase component of the MSAMO. Component B has been further purified by gel filtration and ion-exchange chromatography. A protein complex of around 200 kDa has been identified after gel filtration that resolves into two major polypeptides of approximately 70 and 25 kDa, suggesting a $\alpha_2 \beta_2$ structure. We are speculating at this stage that this may be the hydroxylase component of the MSAMO, due to the multimeric nature of this protein and the identification of specific roles for components A and C.

Further purification to homogeneity, of component C by S-75 Sephadex gel filtration and MonoQ HR5/5 ion exchange chromatography has revealed that this protein is a monomer of approximately 16 kDa. Its N-terminal sequence has significant homology to the ferredoxin components of some dioxygenases (e.g. benzene dioxygenase), and the protein contains two mol of Fe and S per mol. The optical spectrum of component C shows similar characteristics to other Rieske [2Fe-2S-containing] ferredoxins (Higgins *et al.* manuscript in preparation).

The separation of MSAMO into 3 components, all of which are required for restoration of activity *in vitro*, has allowed the analysis of mutants of Strain M2 unable to grow on MSA (MSA^-), isolated after nitrous acid mutagenesis. These do not grow on MSA but still retain the ability to utilize other C_1 compounds (methanol, methylamine and formate) with growth rates similar to that of wild-type Strain M2. These mutants could be grown under derepressing conditions in batch culture in medium containing 20 mM MSA and 6 mM formate, the formate being present to allow sufficient biomass to be produced for analysis. Once all the formate was utilized, the MSA present could induce expression of MSA enzymes in the cells. Oxygen electrode experiments showed that endogenous oxygen consumption rates of mutant cells was not stimulated by MSA although they responded to the addition of other C_1 compounds in the same way as wild-type Strain M2. In some of these MSA^- mutants, MSAMO, activity *in vitro* could be restored by 'biochemical complementation' of cell-free extracts with partially purified Components A, B or C of the MSAMO, suggesting that these mutants may be defective in one or more of the structural polypeptides of the MSAMO (Higgins *et al.* 1995). The isolation of MSA^- mutants will facilitate future studies on the regulation of MSA oxidation in Strain M2 and cloning (by complementation of mutants) of the genes encoding the MSAMO enzyme.

Preliminary examination of cell-free extracts of the marine MSA-utilizers, strains TH3 and PSCH4, have shown that a similar MSAMO enzyme system may be present in these organisms (Thompson *et al.* 1995). The development of antibody and gene probes for this enzyme system should allow further comparisons and perhaps confirm that in gram negative MSA-utilizers, a common mechanism exists for the metabolism of MSA.

Molecular genetics of Strain M2

In order to clone the genes encoding MSAMO and to analyse the regulation of MSA oxidation by Strain M2, it has been necessary to develop a system for mutagenesis of Strain M2 and gene transfer techniques for this organism. Although conventional mutagenesis using nitrosoguanidine resulted in the isolation of mutants of Strain M2 defective in C_1 metabolism, these mutants proved to be unstable. Subsequently we developed nitrous acid mutagenesis techniques (Higgins *et al.* 1995; Davey and Murrell, unpublished) that enables the isolation of stable MSA^- mutants. The development of genetic techniques for Strain M2 has proved to be more difficult since the organism grows slowly and there is a limited range of antibiotic resistant markers that can be used with Strain M2. However, transfer of broad-host range plasmids of the IncP and IncQ groups carrying kanamycin resistance can be successfully transfered to Strain M2 by conjugation (DeMarco and Murrell, unpublished) and experiments are now underway to complement MSA^- mutants using clone banks of chromosomal DNA from Strain M2 constructed using these vectors.

The other strategies being used to identify and isolate genes encoding MSAMO are to utilize N-terminal amino acid sequence information from purified MSAMO-specific polypeptides to design degenerate oligonucleotide probes with which to probe chromosomal DNA from Strain M2. We have successfully used an oligonucleotide probe targeting the 5' end of component C to identify a 6 kbp *Sph*1 chromosomal DNA fragment from the chromosome of Strain M2. This has subsequently been cloned, and partially sequenced. We have confirmed that it contains the gene encoding a Rieske-type [2Fe-2S] ferredoxin and a gene encoding an MSA-specific polypeptide of 45 kDa, previously identified by polyacrylamide gel electrophoresis of cell-free extracts of MSA-grown cells of Strain M2. The interesting possibility that the genes encoding the MSAMO enzyme complex are clustered on the chromosome of Strain M2 is now being investigated by DNA sequencing of this *Sph1* DNA fragment, and comparisons are being made with N-terminal amino acid sequences of purified MSAMO components (DeMarco and Murrell, unpublished). The isolation of MSAMO genes will allow the development of MSA-specific gene probes in order to exploit molecular ecology techniques (see Chapter by Murrell *et al.* this volume) for the study of MSA-degrading bacteria, particularly the marine representatives, in natural environmental samples.

Phylogenetic probes based on the 16S rRNA sequences of MSA-degrading bacteria are also being developed in our laboratory. We have recently completed the sequencing of the 16S rRNA gene from Strain M2. Preliminary phylogenetic analysis has indicated that this may be a novel genus which is phylogenetically closer to *Agrobacterium* than *Methylobacterium* (A. Holmes *et al.* unpublished). 16S rRNA analysis of the two marine strains is also being carried out and it will be interesting to learn if these also belong to a similar genus to Strain M2, since this will allow the development of 16S rRNA probes to detect these bacteria [using fluorescence microscopy techniques (e.g. see Homes *et al.* 1995)] in natural environmental samples.

Conclusions

There is still a considerable amount to learn about the physiology, biochemistry and molecular genetics of MSA utilization in bacteria. Thus far we have only isolated strictly aerobic gram negative representatives of MSA-degraders but there may be a number of different anaerobic/fermentative modes of metabolism for MSA. Recently it has also been shown that a number of bacteria can utilize MSA as a sulfur source (Uria-Nickelsen *et al.* 1993) and it remains to be seen what the role of these organisms in the mineralization of MSA may be. If 'fossil records' of MSA-deposition on the Earth's surface are accurate, then this organic sulfur compound which could serve as a carbon source for aerobic (and anaerobic) methylotrophic growth, and as a sulfur source for many bacteria and possibly fungi, may be mineralized by a much greater diversity of organisms than we presently have in culture.

References

Baker SC, Kelly DP, Murrell JC (1991) Microbial degradation of methanesulphonic acid: a missing link in the biogeochemical sulphur cycle. Nature 350:627-628

Baker SC (1992) PhD thesis, University of Warwick, Coventry, UK

Charlson RJ, Lovelock JE, Andreae MO, Warren SG (1987) Oceanic phytoplankton, atmospheric sulphur cloud albedo and climate. Nature 326:655-661

Higgins TP, Davey M, Trickett J, Kelly DP, Murrell JC (1995) Metabolism of methanesulphonic acid involves a multicomponent monooxygenase enzyme. Microbiology, submitted

Holmes AJ, Owens NJP, Murrell JC (1995) Detection of novel marine methanotrophs using phylogenetic and functional gene probes after methane enrichment. Microbiology 141:1947-1955

Johnston JB, Murray K, Cain RB (1975) Microbial metabolism of aryl sulphonates. A reassessment of colorimetric methods for the determination of sulphite and their use in measuring desulphonation of aryl and alkylbenzene sulphonates. Antonie Leuweenhoek 41:493-511.

Kelly DP, Malin G, Wood AP (1993) Microbial transformations and biogeochemical cycling of one-carbon substrates containing sulphur, nitrogen or halogens. In: Microbial Growth on C_1 Compounds (Murrell JC, Kelly DP, eds) pp 47-64. Intercept, Andover, UK

Kelly DP, Baker SC, Trickett J, Davey M, Murrell JC (1994) Methanesulphonate utilization by a novel methylotrophic bacterium involves an unusual monooxygenase. Microbiology 140:1419-1426

Mason JR, Cammack R (1992) The electron transport proteins of hydroxylating bacterial dioxygenases. Annu Rev Microbiol 46:277-305

Saigne C, Legrand M (1987) Measurements of methanesulphonic acid in Antarctic ice. Nature 330:240-242

Thompson AS, Owens NJP, Murrell JC (1995) Isolation and characterization of methanesulfonic acid degrading bacteria from the marine environment. Appl Env Microbiol 61:2388-2393

Uria-Nickelsen MR, Leadbetter ER, Godchaux W III (1993) Sulfonate utilization by enteric bacteria. J Gen Microbiol 139:203-208

WORKING GROUP 1

Contribution of Microbial Processes to Global Budgets

Bo H Svensson [Rapporteur]
Department of Water and Environmental Studies
University of Linkoping
S-583 81 Linkoping
Sweden

Working Group Participants: P Frenzel, S Jordan, N Iversen, Å Klemedtsson, C Kruse, P Martikainen, D Nedwell, R Oremland, N Panikov, R Segers, P Steudler, A Watson, T Willison

Introduction

Microbial processes are involved in the exchange of most of the trace gases between the atmosphere and most types of ecosystems. Consequently the regulation of the amounts of gases emitted or taken up are tightly linked to changes in microbial activity.

The Working Group discussed different trace gases (excluding carbon dioxide) with the aims of producing an evaluation of our knowledge of (i) the microbiological processes involved, (ii) the magnitude of the resulting fluxes of trace gases, and (iii) the identification of some new or neglected research areas. The gases dealt with were methane, nitrous oxide, CO, DMS, COS, CS_2 and chlorofluorocarbons (CFCs). For some of these gases the net flows are fairly well known for different types of ecosystems, although depending on their inherent spatial and temporal heterogeneity, the global fluxes show large variance. In order to narrow these ranges, data lending a stronger coupling between the regulation of the microbial processes and the fluxes need to be sought. Such data should allow for the construction of mechanistic models acting as modulators on flux rates in relation to changes in climate and soil conditions. For some of the gases and ecosystems, the

NATO ASI Series, Vol. I 39
Microbiology of Atmospheric Trace Gases
Edited by J. Colin Murrell and Donovan P. Kelly

formulation of such models is now under way (e.g. for CH_4 in mire ecosystems).

Methane

More than 80% of the net global methane flow to the atmosphere (~535 Tg/year) is of modern microbial origin. A little less than two-thirds of the microbially-generated methane derives from anthropogenic activities such as waste management systems, stock breeding and rice paddies, while the rest comes from natural sources, such as wetlands, freshwater and marine habitats, and termites. Several of these source systems harbour methane-oxidizing microorganisms, which intercept some of the methane emission. The microbial production of methane in these source systems is thus greater than the amount emitted to the atmosphere. Dry soils generally act as a sink for atmospheric methane. The estimated net removal of methane by microbial oxidation in dry soils amounts to 5-7% of the estimated net emission. It should be noted that this amount corresponds to the estimated annual burden on the atmosphere giving rise to the annual increase in methane concentration of the atmosphere.

The main microbial processes (i.e. methanogenesis and methanotrophy) are quite well understood on the cellular and ecophysiological levels. Anaerobic oxidation of methane is, however, observed in marine sediments and hypersaline inland waters, and seems to remove substantial amounts of the methane generated in these ecosystems. The organisms responsible for this are unknown, which poses a considerable (and as yet unresolved) research challenge. A promising development is that recently, strong evidence has been presented to support the view that methanogenic bacteria oxidize methane in a syntrophic association with sulfate-reducing bacteria as [H]-consumers.

The consumption of methane at atmospheric concentrations by the top soils of most terrestrial ecosystems is not yet attributable to known organisms, and until the organisms or consortia responsible are characterized physiologically it will not be possible to explain the regulation of soil methane oxidation. Ammonia-oxidizing (nitrifying) bacteria can also oxidize methane, and work is required to understand the coupling between methane and ammonia oxidation and its relationship to observed nitrifying activities in the same soils. Establishment of which bacteria catalyze methane destruction, operating individually or as consortia, is a vital part of understanding the soil as a sink for methane.

While it is well-established that the increased methane burden in the atmosphere is largely attributable to human activities, we have as yet very little evidence concerning any effects on the microbial sink activities for atmospheric methane, resulting from the rise in ambient methane levels. Any such positive changes might either enhance or mitigate the net emissions to the atmosphere, and research into elucidation of the regulation of methane consumption in terrestrial ecosystems is obviously of high importance. Areas undergoing changes in land use, and areas affected by climatic change, should be targeted for such studies.

Nitrous oxide

On a global basis the estimated annual net emissions of N_2O (11.5 Tg N) are believed to be 70% from natural sources (soils and oceans) and 30% anthropogenic. The stratospheric sink for N_2O is estimated to be 12.3 Tg N/year, but there is an annual *increase* in the atmospheric N_2O burden of 3.9 Tg N. This means that there is a shortfall in the known sources amounting to 4.7 Tg N/year. No account has previously been taken of a possible soil sink, which would further enhance the requirement of a higher source strength. Microbial processes account for about 85% of the source output, of which about one-fifth results from human stimulation of activity (soil cultivation, cattle, and cattle feed lots).

The most important microbial processes giving rise to N_2O are nitrification and denitrification. Denitrifying bacteria are able to consume N_2O and may therefore act as regulators on N_2O pools in water and soils. Since at least these two microbial processes exist for N_2O production, and each exhibits very different regulatory patterns for their respective N_2O-forming systems, it is not surprising that great variation in emission rates is observed in natural ecosystems. The nitrification and denitrification processes are often active in the same horizon or aggregate of a soil, or conceivably on the same suspended particle in the ocean. This makes very difficult the pinpointing of a major regulator in the net flow of N_2O, or the construction of simple general models as tools to help generate flux data with greater confidence. Interactions with plant roots further complicate understanding of these processes.

Despite the enormous amount of data available on the biochemistry of nitrification and denitrification, considerably more research on the ecophysiology and *in situ* activity of bacteria producing and consuming N_2O is required in relation to the flux of this very important gas to the atmosphere.

At this meeting, the acetylene inhibition technique was brought into question. It has been considered an excellent tool in studying the nitrifiers, but was here reported to give severe underestimation of the denitrifying activity of systems in contact with water. This means that many of the laboratory and field incubation studies carried out over the past 20 years, and used to give boundary conditions and potential rates of denitrification, may be misleading. New techniques to study denitrification should therefore be developed to define the regulatory effects on these bacteria in pure cultures and in natural systems.

Carbon monoxide

CO is formed by some soil organisms, but has a major oceanic source and anthropogenic sources. Globally, soil is considered as a CO-sink and is estimated to sequester 10-20% of the annual CO flux of 2-3 x 10^3 Tg.

Dimethyl sulfide, carbonyl sulfide and carbon disulfide

Emissions of all these sulfur gases are known to occur in forests, open and coastal marine environments, and wetlands. No data are available from grasslands, deserts and tundra. DMS emissions have been observed also from arable land and DMS and CS_2 also arise from fresh waters. The absolute emission values for these gases are known only with considerable uncertainty, but maximal annual emissions are estimated at 45, 3, and 5 Tg each for DMS. COS and CS_2. Recent estimates put the output of DMS from the oceans at 20 Tg S/year, with total global output at 22.6 Tg S/year. All other biogenic sulfur gases (including CS_2, COS, H_2S and dimethyl disulfide) probably total 23.7 Tg S/year.

A substantial part of the DMS formed by microorganisms in oceans and wetlands is subjected to microbial turnover. The fate of DMS has been studied in depth during recent years. It is likely that earlier estimates of DMS output are too high, as it is known that dimethylsulfoniopropionate (DMSP) also gives rise to methanethiol (possibly more than 50% DMSP enters this route).

CS_2 arises from cysteine and isothiocyanates in soils, and is known to be oxidized for growth of some *Thiobacillus* strains. COS arises photochemically from organic precursors in the oceans and from atmospheric oxidation of CS_2, but its biogenic progenitors are much less well known. Ultimately these compounds return to the Earth in precipitation, contributing sulfur (as sulfate or methanesulfonate) to the biosphere. Their climatic importance has yet to be fully evaluated.

CFCs

Some of these anthropogenic gases can be metabolized by microorganisms, but the concentrations of these materials are so low that microbial scavenging from the atmosphere seems improbable. Higher concentrations have been observed in landfills, and some CFCs have been shown to undergo transformation when incubated with anaerobic wastes: this merits more microbiological research as a potential means of mitigating CFC inputs to the atmosphere from anthropogenic wastes in landfills.

WORKING GROUP 2

Global Environmental Change (GEC)

Donovan P Kelly [Rapporteur]

Institute of Education
University of Warwick
Coventry CV4 7AL
England

Working Group Participants: O van Cleemput, R Conrad, A Costello, S Jensen, L Kargiolaki, G King, N Kubilay, M Lidstrom, P Megonigal, R Oremland, P Steudler.

Trace gases and global environmental change

Atmospheric trace gases exert one or more of a series of possible effects on climate homeostasis, potentially leading to climate change or altered radiative input to the biosphere. Most important of these are:

1. *"Greenhouse" effects*, caused by gases reacting with input solar radiation to retain heat in the atmosphere, and thus raising mean global temperature. Most important are water vapour and carbon dioxide, primarily because of their abundance in the atmosphere (and currently accounting for at least 85% of the warming effect of the sun in maintaining the Earth at a mean temperature some 30°C higher than the -18°C it would enjoy in the absence of an atmosphere). There has, however, been an increase in mean temperature due to radiative forcing by other trace gases. Over the past 15 years this forcing has totalled around 550 mW/m^2/decade, and is attributed to carbon dioxide (~56%), methane (~15%, including that due to water vapour from

NATO ASI Series, Vol. I 39
Microbiology of Atmospheric Trace Gases
Edited by J. Colin Murrell and Donovan P. Kelly

atmospheric oxidation of CH_4 by OH), nitrous oxide (~6%) and anthropogenic chlorofluorocarbons (CFCs; ~24%).

2. *Ozone-depleting gases*. Most significant among these are the CFCs and biogenic chlorine and bromine compounds, which persist in the atmosphere long enough to penetrate to the stratosphere and react in the ozone layer. The CFCs thus exert a dual role, but it has to be noted that stratospheric ozone is also a "greenhouse gas", and there is some evidence suggesting that ozone destruction in the Antarctic partly offsets global warming due to CFCs in the Northern hemisphere.

3. *Cloud cover and water condensation*. Some trace gases (or their atmospheric oxidation products, such as methanesulfonate from DMS) act as cloud condensation nuclei, potentially resulting in droplet formation, increased cloud cover, and enhanced precipitation. They could thus counter the increased atmospheric warming resulting from a greater atmospheric burden of water vapour, and potentially reduce radiative forcing by increasing global albedo.

Modelling change and trace gas dynamics

Major factors modelled in short- to medium-term considerations of GEC are (i) atmospheric CO_2, and its effect on atmospheric temperature and on plant growth, (ii) temperature change and its consequences, (iii) depositional change, and (iv) global water relations. To these, the work of the past decades in atmospheric chemistry and physics and in microbiology has added (v) methane, nitrogen oxides, organosulfur gases, the chloro- and chlorofluoro-carbons, and diverse other trace compounds containing nitrogen or halogens.

The trace gases discussed in the ARW affect and influence, or are influenced by, all these factors. Two aspects of microbiology are relevant: (i) the effect of variation in trace gas production through microbiological processes on possible GEC; and (ii) the prediction of effects of GEC on trace gas dynamics as driven by microbiological processes. These interrelated aspects depend on having knowledge of the bacteria and communities involved, and the numbers, functions, activities and potentials of these organisms.

Two blocks of inter-related studies need to inform a consideration of trace gases in GEC. First, the range, nature, sources and sinks, and atmospheric effects of the trace gases themselves; and secondly, the physical, chemical and anthropogenic influences on those gases and their dynamics.

The gases of interest are methane, sulfur gases (dimethyl sulfide, carbonyl sulfide), nitrogen gases (NO, N_2O, NO_x), and halogenated gases primarily of biogenic (Cl, Br) or anthropogenic (Cl, F) origins.

Among the principal factors affecting dynamics, both of sources and sinks and of atmospheric turnover, are (i) pCO_2, (ii) temperature, (iii) hydrology, (iv) N-deposition, (v) land use change (anthropocentric), (vi) non-biological source strengths, (vii) physiological diversity and interactions between microbial communities with internal element cycling having variable freedom of exchange of intermediates (which includes the trace gas output) with the adjacent environment and the atmosphere, and (viii) variability, unknown and unquantified sinks.

We are currently very patchy in our understanding of the interrelations and controls over trace gas outputs and the "feedback" effects (both positive and negative) that affect the fluxes of trace gases between their sources and sinks. The discussions in this Working Group concentrated on defining the major factors affecting processes and ensuring that these received priority recognition by experimentalists in field microbiology.

Modelling feedback loops and homeostasis

In principle it is possible to correlate all the factors influencing trace gas turnover (output from soils or aquatic sources, sinks in the atmosphere and the biosphere, interactions with physical and anthropocentric processes) to a set of differential equations, as the various factors can each be defined with a greater or lesser degree of precision. Thus, change in methane input to the atmosphere with respect to change in temperature could be expressed in relation to the changes due to altered nitrogen deposition, soil disturbance, altered wetting, water activity (a_w), and so on, e.g.:

$$dCH_4/dt = [\partial CH_4/\partial N_{dep} + \partial CH_4/\partial T + \partial CH_4/\partial a_w + \partial CH_4/\partial_{toxins} + \partial ...]$$

Clearly each component expression would be subject to variable uncertainty: for example, the change of CH_4 with temperature would be a function of the effect of temperature on bacterial methanogenesis and methanotrophy (by both methanotrophs and ammonia-oxidizers). These contributory processes could also be directly or indirectly affected by altered water activity or aeration (due to land disturbance or altered hydrology), making the $\partial CH_4/\partial T$ function a combined physical (solubility and diffusion) and biochemical variable. Nonetheless it should be an aim of the biological and physical studies of trace gases and their turnover to identify the principal numerical components needed in order to express gross atmospheric input and output in such terms. Only in this way can the interrelations of the major physical and biological variables on the cumulative effect of diverse gases be quantified, however crudely.

Effects of physicochemical change on trace gas turnover by the biosphere

Increases in pCO_2 and temperature both potentially increase trace gas turnover and flux of gases like methane and nitrous oxide to the atmosphere. Raised CO_2 would be expected to enhance primary production (especially of C_3 plants), thereby making more carbon available for methanogenesis and nitrate reduction. By enhancing biological activity, temperature rise could stimulate all gas-producing (and consuming) processes. Temperature could, however, also affect water availability (and the water table in methane-producing wetlands for example) and also thereby exert a negative effect both by reducing the volume of the habitat and through influencing the water potential of a system.

Globally it is likely that the Northern wetlands will decrease in area both by human activity and by hydrological changes resulting in drying. This would reduce methanogenesis, but a counteractive activity is likely to be the expansion of methane-producing rice paddies in Asia over the next half century, and possible increases in CH_4 arising from increased populations of ruminants in the tropics. Increased drying due to temperature rise might decrease the methane-productive area of the Arctic wetlands but it seems likely that these regions also become less effective as CH_4 sinks if temperature is raised. This seems to be due to decreased methanotrophy, while methanogenesis is enhanced. Modelling the effect of temperature and drying on wetland methanogenesis is thus complex and requires more data.

The effect of human land use changes is also variably predictable. Draining and enhancing land aeration would be expected to reduce methanogenesis and nitrous oxide output, but drainage of wetland has been seen to result initially in enhanced output of N_2O, possibly attributable to exposure of more carbon to organisms capable of denitrification.

Interactions between GEC and trace gas dynamics: a model for methane

The effects of the factors, ∂pCO_2, ∂T, $\partial N_{deposition}$, $\partial N_{fertilization}$, $\partial_{land\ use}$, $\partial_{hydrology}$ on the sources and sinks for methane were considered by the Working Group in drawing up a scheme of relative importance of the factors to the source and sink terms.

1. Source terms

Wetlands: Elevated pCO_2 would increase primary production by C3 plants in wetlands, thereby increasing all microbial processes involved in the mineralization of organic matter. This effect could be enhanced by nitrogen deposition or eutrophication, and amplified by increase in mean temperature.

Elevated temperature will generally enhance microbial metabolism (Q_{10} values may exceed 2), but also decrease gas solubilities. Lowered oxygen solubility could possibly reduce the rate of CH_4 oxidation, but methanogenesis could increase, with decreased solubility also meaning that ebullition of dissolved CH_4 could be raised.

Change in water status (+ or -) would be likely to have an equivalent effect on methanogenesis, as the volume of the methanogenic habitat varied. This would be time-dependent, as wetland plant (and hence carbon input) changes would require successional seres, and carbon and mineral availability might be rate-limiting to methanogenesis.

Increased nitrogen input by atmospheric deposition could increase primary production, stimulating methanogenesis. Similarly ingress of N from agricultural sources would have an effect that was interdependent with changes in pCO_2, temperature and water.

Land use changes resulting in decreased wetland would enhance CH_4 oxidation and decrease observed CH_4 output.

Rice paddies: Elevated pCO_2 might increase the rice primary production, and root-associated methanogenesis could be stimulated if recent photosynthate is a major input for methanogenic carbon. If below-ground biomass from previous rice culture (and any added organic matter) are major sources, the effect of pCO_2 would be delayed until accumulated biomass increased seasonally.

Effects of other parameters would be similar for other wetlands, although atmospheric N deposition would have at best a marginal effect.

Ruminants: Changes in pCO_2, temperature etc. would not have direct effects on ruminant methanogenesis, but this could be affected by land use and fodder possibly consequent on those other changes.

Termites: GEC might affect habitat distribution (geographical areas suitable for colonization), and consequently the regions producing methane from termites. Any changes affecting the availability of plant fibre (wood) available for termite nutrition (e.g. ∂T, $\partial N_{deposition}$, hydrology) would influence CH_4 output, and increased temperature could raise CH_4 production simply by enhancing activity and metabolism.

Landfills. In that these are man-managed, only directed changes in drainage, area of land used, and hydrologic regime would affect them, and changes in pCO_2 would not have any effect. Any temperature enhancement would probably have only marginal effects on a large, self-heating, landfill but any gross increase might enhance methanogenesis and hence CH_4 venting, but an increase in surface temperature, possibly accompanied by consequent drying, might decrease methanotrophy, thereby allowing greater CH_4 input to the atmosphere.

Oceans: Coastal eutrophication might increase near-shore methanogenesis and sea-level rise could alter the areas of coastal marshes from which CH_4 is emitted. No significant effect on oceanic CH_4 generation would be expected, although surface temperature rise could increase the rate of loss from the ocean to the atmosphere.

2. Sink terms

Soils: pCO_2 would have no direct effect on CH_4 consumption in soils, except indirectly if ammonia-oxidizing autotrophs were to become more abundant and effect enhanced CH_4 destruction. Temperature would have little direct effect as atmospheric methane consumption by soils is diffusion-limited. Enhanced oxidation might accompany enhancement of methane transport resulting from decreased soil moisture at elevated temperature.

Nitrogen deposition from the atmosphere can result in long-term decreases in CH_4 oxidation, and changes in the relative importance of the activities of methanotrophic and ammonia-oxidizing bacteria. The magnitude of the nitrogen input effects is likely to be a function of soil type (texture) and the *in situ* nitrogen mineralization and nitrification potentials. Losses of forests, prairies or other land types (e.g. conversion to agricultural use) and intensive forest management all result in long-term decreases in CH_4 oxidation, and potential loss of sinks for this gas.

Other methane sinks: Methane oxidation in wetlands appears primarily to be a function of oxygen availability. Decreases in the water table levels increases availability, while higher temperature and greater organic content reduce dissolved oxygen supply. One would expect latitudinal variation in wetland methanotrophy consequent on GEC.

The factors considered would have minimal effects on the oxidation of CH_4 in rice paddies, landfills or the oceans, although the value of the two former as sinks could be enhanced by appropriate human management. Little methane oxidation appears to be associated with ruminants or termites, although oxidation in the soil around the latter would obviously be influenced by ∂T, ∂_{water}, and ∂N.

The analysis given above is of necessity superficial and serves to show the kind of effects that would apply to biogenic gases other than methane. Thus, plant-produced gases (carbon disulfide, volatile acids and esters, DMS) could be affected by pCO_2, temperature and nutrient status (e.g. iron or phosphate for marine sources), but many of the principles are generally applicable. For most biogenic trace gas sources we have little secure knowledge of the stability of the source ecosystems and the main drivers of any homeostasis they exhibit. Systems such as the DMS-producing oceans are of very uncertain "stability". It is known that much DMS never leaves the ocean surface, as it is reoxidized and recycled within the ocean ecosystem: this may account for as much as 90% of all biogenic DMS. Consequently temperature and nutrient effects on the oceanic DMS source (and consequent sulfur input to the atmosphere) are impossible to model at this time. Indeed, we have to admit to an overall lack of predictive understanding of the correlations between CO_2 production, pCO_2, temperature and sink regulation in the modelling of atmospheric inputs of the major trace gases.

Other factors affecting gas cycling

Some processes are subject to marked chemical limitations. Thus, the oxidation of NO to NO_2 ($2\ NO + O_2 = 2\ NO_2$) is dependent on two NO molecules reacting with O_2 and is consequently slow when NO concentrations are low. This means that below 10 ppm, NO exhibits an atmospheric lifetime exceeding 100 h, and oxidation of concentrations in the ppb to 1 ppm range in soils will be a balance of chemical and biological processes, dependent on the [NO]. Also, the proportions of DMS oxidized methylotrophically, phototrophically and venting to the atmosphere from the ocean varies with latitude: biological oxidation in the tropical ocean appears maximal near the equator, with all three options being equivalent at 12°S.

The effects of toxic materials and the interdependence of processes involving different trace gases can influence output rates and steady state levels. Thus a gas such as methyl fluoride (and some CFCs) can be inhibitory to methanotrophy, but inhibition by the former can be overcome as the "toxic" material is also a substrate and will eventually be oxidized. The processes of denitrification, nitrification and methanotrophy show overlaps in physiology and the first can be linked to the anaerobic destruction of methylamines and sulfides.

Other element cycling

Although the ARW (and this Working Group) focused primarily on the atmospheric aspects of global change as catalyzed by microorganisms, a further dimension was recognized that affects environmental quality. This concerned the involvement of microorganisms in the global cycling of toxic elements. as exemplified by mercury, and to a lesser extent elements such as selenium and arsenic, all of which undergo volatilization *via* microbial methylation. Broadly-based anthropogenic activities, including fossil fuel combustion, hydrological alterations, and other industrial operations have mobilized these elements and increased their flux to the atmosphere from various point sources. Their fall-out from the atmosphere has resulted in large-scale contamination problems in a number of ecosystems. The involvement of microbes in their immobilization, volatilization, and detoxification is relatively poorly understood and merits further quantitative and biochemical assessment because of its significance to plant and animal health, and as a factor in trace gas cycling that might affect the activities of bacteria whose normal metabolism, for example, uses non-toxic methylated compounds.

Future work

Predicting the effects of climate change on the trace gas dynamics driven by microorganisms (and the consequent further change driven by any modification of those microbial dynamics) depends on having knowledge of the bacterial types involved, community interactions, and *in situ* numbers, functions and activities. This requirement cannot yet be fully met: the development of methodologies still lags behind need. In particular the development of techniques is needed that will establish the identity, physiological capabilities, and reaction velocities (for target processes relating to trace gas turnover) for the 95% of all natural microbial populations which have as yet resisted attempts to culture them in the laboratory.

Acknowledgements

The rapporteur for this Working Group is grateful to Gary King and Ron Oremland for inputs provided during the ARW, which have been incorporated into the text of this report.

WORKING GROUP 3

Problems and Developments in Methodology

Andrew Holmes[1] and David Lloyd[2] [Rapporteurs]

[1]Department of Biological Sciences
University of Warwick
Coventry CV4 7AL
UK

[2]Pure and Applied Biology
University of Wales College of Cardiff
PO Box 915
Cardiff CF2 1TL
UK

Working Group Participants: L Bodrossy, R Conrad, A Costello, K Henricksen, S Jensen, E Kenna, M Lidstrom, W Liesack, I McDonald, C Murrell, A Nielsen, N Revsbech, P Roslev, M Slayne, D Smart, J Sorensen, T Willison.

Microbiology of trace gases

Global change, and in particular the Greenhouse effect, has become a key issue in recent times. The significance of a particular gas in the Greenhouse effect is proportional to both its atmospheric concentration and its potential for thermal absorption. The trace gases CO_2, CH_4, and N_2O all have long atmospheric residence times (5-164 years) and are major agents of global change. For this reason in developing models to predict climate change due to atmospheric trace gases, the underlying chemical and biochemical processes controlling the emissions must be understood. These processes are carried out principally by micro-organisms. This report is a condensation of discussion on methodological problems and developments in the investigation of the microbiology and biochemistry of trace gas emission.

NATO ASI Series, Vol. I 39
Microbiology of Atmospheric Trace Gases
Edited by J. Colin Murrell and Donovan P. Kelly

Two approaches to study of the microbiology of trace gases have been developed; (i) microbially-mediated processes can be looked at in general terms by examining the effect of environmental perturbations on gas emission/deposition and thereby elucidate biochemical mechanisms; or (ii) population structure can be analyzed to determine which microbial groups are important in regulation of fluxes. The ultimate goal is to obtain accurate measurements, on many different scales, that can be used in construction of models for prediction of the magnitude of gaseous fluxes both regionally and globally. This can not be achieved solely by either of these approaches but requires a combination of the two. The most problematic area of research lies in developing methodology capable of linking these two approaches together. Therefore the relationships between measurable population structure and gas fluxes must be simplified sufficiently to allow scale-up to regional or global models which still have a reasonable predictive power. We attempted to evaluate methods and pinpoint problems and future developments in the light of these considerations.

Analysis of microbial populations

The enumeration and identification of bacteria is one of the most difficult tasks facing microbiologists. It is now well-characterised that the majority of bacteria in the natural environment are not represented in culture collections and may well be resistant to isolation by present techniques. It follows from this that traditional plate count and MPN methods for enumeration of bacterial groups are unreliable at best and that standard laboratory biochemical tests for identification are not applicable for most bacteria. In the past 10 years the application of molecular biology techniques to microbial ecology has meant that it is now possible to detect uncultivated organisms and identify them with respect to sequence databases.

Methods employed in these approaches revolve around the extraction of constituents (usually nucleic acids) from bulk environmental samples. This mixture of nucleic acids is then separated by cloning to resolve the original members of the population. Species can be identified by sequence analysis of clones and comparison with a database of known species. Whilst this type of approach has solved the problem of detection of uncultivated organisms it has two major limitations; (i) identification is based solely on the sequence of a single gene (in most cases that of the small subunit rRNA) from which the physiology of the organism cannot always be reliably inferred; (ii) these techniques are very expensive and labour intensive.

The most laborious and expensive part of nucleic acid approaches is the construction and analysis of gene libraries. Other methods for resolving the community structure revolve around hybridization of group-specific probes to detect particular organisms within the mixture. Such approaches reduce the cost and time involved in the analysis of an environmental sample by several orders of magnitude, but rely on prior knowledge of an organism's sequence to enable design of suitable probes. For organisms for which probes are available a variety of methods are now available for both enumeration and single-cell identification.

Probes may be applied directly to filter-bound nucleic acids in Southern or Northern dot blots. If binding of nucleic acids to filters is carefully controlled these techniques offer a rapid quantitative analysis of an organisms relative abundance in a sample. Unfortunately this approach is not very sensitive and hence inappropriate for organisms of low abundance.

The sensitivity of detection can be dramatically increased by use of the polymerase chain reaction. The power of PCR is such that less than 10 cells may be detected within a sample making possible investigation of smaller, better defined samples. PCR may also yield quantitative information, but must be carefully controlled as the reaction is sensitive to variations in sample purity which occur in nucleic acids extracted from the environment. PCR may also be of use in emerging methods for community analysis which avoid laborious gene library techniques such as denaturing gradient gel electrophoresis (DGGE) which allows the resolution of double-stranded nucleic acids according to their sequence. Analysis of PCR products from environmental DNA samples by DGGE can give a quick overview of population structure if group-specific probes can be used to identify the source organism for bands resolved on the gel.

Probes can also target sites within whole cells rather than free nucleic acid targets. Cells labelled in this way may be detected by either microscopy or flow cytometry. Another advantage of whole cell hybridization is that it is now possible, for the first time, to identify micro-organisms `in situ'. The major drawback to whole-cell hybridization methods at present is sensitivity. For fluorescently-labelled probes only target molecules which are abundant within the cell (*e.g.* ribosomal RNA's in actively growing cells) give a strong enough signal to allow direct detection by fluorescence microscopy. Sensitivity can be improved by using enzymatically labelled probes but this requires additional steps to permeabilize cells and is not generally applicable to all cell types. A recent alternative approach to increase sensitivity is the use of laser scanning confocal microscopy (LSCM). The scanned focused excitation beam reduces background fluorescence considerably thereby allowing more sensitive light detection. In conjunction with LSCM, fluorescently labelled probes can be used to detect even cells with low numbers of target molecules. LSCM also offers a capacity for optical sectioning allowing high spatial resolution in three dimensions of intact samples.

In all applications of probe technology the principal limitation is the database from which to design probes. The most extensive database is for 16S rRNA genes and at present this is the only molecule suitable for total population analysis. The major problem facing microbial ecologists is that one can seldom assign a physiological or ecological role to an organism identified solely on the basis of 16S rRNA.

Measurement of microbially-mediated processes

The most useful predictions of microbially-mediated response of regional flux to environmental change come from physiological data on the causative organisms. This area, identification of microbial regulatory factors of gas fluxes, presents the most pressing methodological problems and is one which will increasingly require interdisciplinary approaches.

At present methods for directly measuring microbially mediated processes do not allow activities to be ascribed to a particular group of bacteria. The most useful techniques are those involving specific inhibitors. Gas chamber analysis allows the flux of gases to be accurately measured in a controlled environment. Use of specific inhibitors allows blocking of certain steps in gas formation or decomposition permitting calculation of the contribution of these stages to the net flux. Their use must be carefully controlled to allow interpretation of results, particularly with respect to the specificity of inhibition. In cases where the specificity of the inhibitor and its `behaviour' in the environment (*e.g.* solubility, diffusion time, and half-life) are well-understood this is the method of choice. A major concern raised was the validity of laboratory cultures as reference strains for use of specific inhibitors. In many cases there is evidence that lab cultures contain different isoforms of the enzymes to naturally occurring strains so these may respond differently to an inhibitor compound. As new isolates are obtained and better understanding of the biochemistry reached, more efficient inhibitors may be discovered.

Isotopic tracers may also be used in a similar manner to specific inhibitors for the identification of sources and sinks for a compound. In some cases the isotope may be specifically incorporated into cell biomass allowing direct identification of the key organisms. In most cases however, it is not possible to identify the organisms responsible for a process using these methods despite their capacity to measure the net process.

One approach to identifying the organisms is differential rate comparisons. Steudler and colleagues (this volume) have exploited the different kinetics for CO and CH_4 oxidation by methanotrophs and nitrifiers to identify which of these groups is responsible for methane oxidation in soils. This strategy offers a useful means for indirect identification of key organisms but is reliant on the relevance of laboratory reference strains as "kinetic standards". As discussed above this may not always be the case.

As discussed above the methodology now exists for detection of essentially all micro-organisms and to some extent different groups can be enumerated. However, identification of physiological groups by these methods is only possible if representatives are available in pure culture. It is therefore difficult to correlate data on physiological processes with data obtained from microbial populations.

This problem is further compounded by the difficulty in distinguishing between the active and dormant fractions of a population. Several workers have demonstrated a correlation between growth rate and ribosomal content which has led to attempts to determine growth rates of natural populations using whole-cell hybridization to rRNA-targeted probes. Whilst potentially very useful it is not clear at present if inferral of growth rate by this method will be possible for all bacteria. A further difficulty is that growth rate seldom correlates with activity and is therefore not a valid indicator of a cells contribution to overall rate.

A better indicator of activity may be transcriptional rates, or mRNA, levels for genes central to the process being studied. Recent reports indicate it is possible to detect mRNA in whole cells both in laboratory systems and environmental samples. As the genes specifically associated with key biochemical steps are identified, and the databases for these increase, hybridization of "functional group-specific" probes to mRNA in dot blots or whole cells will become increasingly important in detection and estimation of the activity of physiological groups of bacteria.

An alternate technique for determining cell viability is the use of membrane potential sensitive dyes such as rhodamine 123, or cyanins. These dyes may distinguish between metabolically active cells and dormant or dead cells but also cannot directly measure a cells contribution to a particular process.

It is evident from the problems discussed above that the limiting factors for identification of microbial regulatory factors of gas fluxes center around our lack of knowledge of the organisms, their physiological states, and their enzyme activities involved in the regulatory processes for trace gas flux. Clearly non-invasive methods are needed for *in situ* studies which parallel the identification and isolation of both functional groups of organisms and chemical intermediates in gas formation/decomposition. Interdisciplinary studies using the methods described here may facilitate this.

Sampling techniques

Important considerations in developing models to describe a regional gas flux are the scales and extents of sampling needed to be representative. As has been discussed detailed analyses of population structure is time consuming, expensive and will be difficult to interpret. If suitable group-specific probes are available then multiple sampling for detection of specific organisms is possible.

Point fluxes can vary considerably over a region (*e.g.* hummocks vs. hollows in peat bogs). This suggests the activities and perhaps populations of bacteria also vary within a region. To obtain a representative sample of the microbial population or its activity samples should thus encompass both a representative range of microhabitats and climatic variables must be considered.

For detailed understanding of the microbiology of trace gas fluxes the invasiveness of sampling techniques is often a limiting factor. The capacity for spatial resolution and continuous sampling in dynamic environments are both severely affected by this. The recent development of microprobes with 1μM detection limits and outer diameter as small as 10μm represent a great advance in our ability for non-invasive measurements. The portability and price of such equipment makes spatial resolution of processes in field samples feasible. Correlation of these processes to activity of particular populations is a less tractable problem. Techniques requiring nucleic acid extraction are destructive and difficult to apply to extremely small samples. Whole cell detection methods require samples to be transported back to the laboratory but do offer spatial resolution of bacteria within the sample. LSCM in particular represents a valuable tool, although can only be used for dead, fixed samples and not for living cells. Non-invasive analytical methods which allow real time multiple measurements of a microbial population (*e.g.* NMR) have not been applied to ecological problems, mainly because of inadequate sensitivity.

In many cases workers are attempting to identify which bacteria are responsible for a process. In such cases the choice of sampling site may be just as important as the methods chosen. "Hot spots" should be identified by flux or activity measurements before detailed analysis of populations is undertaken.

Summary

16S rRNA gene libraries allow essentially complete characterization of any microbial community. Limited information on the physiology of new uncultivated species detected in this manner can be inferred by phylogenetic analysis of the sequence information. This technique is unsuitable for multiple sampling or application to large areas so sample sites must be chosen judiciously.

The use of oligonucleotide probes to directly detect specific groups of bacteria is less expensive and quicker than gene library analysis but is restricted by the database from which to design probes. Such probes may be "phylogenetic group-specific" or "functional group-specific". The most extensive database at present is for the 16S rRNA allowing design of phylogenetic group-specific probes. As sequence information becomes available for functional genes design of probes targeting these will become possible. Both such probes may lack sensitivity when applied directly to environmental samples.

PCR increases the sensitivity of nucleic acid based detection methods by several orders of magnitude and gives a greater capacity for throughput of samples. In conjunction with DGGE, PCR may also allow more rapid analysis of community structure than gene libraries.

The application of probes to whole cell hybridization allows more accurate quantitation than PCR-based methods and improved spatial resolution. For rRNA-targeted probes LSCM offers sufficient sensitivity to detect most living bacterial cells. "Functional group-specific" probes targeting other genes (or their mRNA) may not always show sufficient sensitivity.

Whilst nucleic acid based approaches allow the detection of the majority of organisms the only cases in which we can identify a physiologically defined bacterial using this technology is where a representative of that taxon has previously been characterised in pure culture. Either phylogenetic group-specific or functional group-specific probes may then be used. A major task for microbiologists is the isolation and characterization of new species to extend the databases. Of even greater use will be the identification of genes specific for functional groups (*e.g.* methane monooxygenase in methanotrophs) and accumulation of sequence information for these.

The use of increasingly powerful physical techniques promises greater and greater speed and sensitivity of gas and intermediary metabolite monitoring *in situ* over extended periods and with minimal perturbation. Imaging techniques currently used are based on LSCM, but here again developments in instrumentation suggest that a new generation of devices will soon be useful in resolving spatial heterogeneity in structured microbial ecosystems.

WORKING GROUP 4

Sulfur, Carbon and Nitrogen Interactions

Ann P Wood [Rapporteur]

Division of Life Sciences
King's College London
Campden Hill Road
London W8 7AH
England

Working Group Participants: S van Bergeijk, T Hansen, S Joye, R Kiene, G Malin, JCM Scholten, R Simo, P Visscher.

Background

There is considerable variation in the amount and quality of the information concerning these interactions and transformations: for example, relatively less is known about the microbial processes effecting the cycling of sulfur gases compared to the microbiology and biochemistry of methane turnover. There are numerous microbiological and functional overlaps between the cycles of these elements, for example, where one organism type has more than one degradative ability. Thus, some methylotrophs are able to metabolize sulfur-containing compounds such as methanethiol, dimethylsulfide and methanesulfonate, as well as methylotrophic substrates based only on carbon (methanol or formaldehyde) or on carbon-nitrogen compounds (e.g. the methylamines). Natural communities normally contain a diversity of types of organisms, providing metabolic overlap and possible commensal interaction, potentially including co-metabolism of some of these compounds. Overlaps in the microbiological transformations of sulfur compounds with reactions of the nitrogen cycle is in a way analogous to the

NATO ASI Series, Vol. I 39
Microbiology of Atmospheric Trace Gases
Edited by J. Colin Murrell and Donovan P. Kelly

methane/ammonia-nitrogen overlap. Links between the sulfur and nitrogen cycles are most obviously seen to be mediated by the methylated amines and the methylated sulfur gases.

Key issues

The main interest in this area was identified as the ecological relationships between the organisms effecting the degradation of these compounds. Intensive research on these processes and interactions is necessary because of the significance of many of these gases and their turnover to environmental change processes. The areas of interest of this working group encompassed a multitude of microbial types active in a wide range of environments, principally transforming mixed carbon-sulfur or carbon-nitrogen substrates, and it was primarily these topics which were the subject of further discussion.

Methylated sulfur and nitrogen compounds: dimethylsulfide (DMS), dimethylsulfoniopropionate (DMSP), methanethiol (MT) and methylamines

These compounds represent principal links between the sulfur, nitrogen and carbon cycles. Their roles are especially noticeable in algal blooms where compounds such as DMS and DMSP may be the major part of a carbon pool: for example, DMSP can account for 1-8% of phytoplankton-carbon. Phototrophs such as *Spartina alterniflora*, *S. foliosa* and *S. angelica* produce DMSP but not glycine betaine which is found in many other organisms (e.g. invertebrates, fish, sugar beet and humans). While the relative affinities of bacteria for DMSP rather than glycine betaine or other organic compounds released into the ocean is uncertain, they may be important during bloom turnover.

The availability of these compounds in natural environments (for comparison with conclusions drawn from *in vitro* experimental conditions) is not globally known but is heterogeneous at best. Localized high concentrations of some compounds occur, such as DMSP and DMS when algal lysis occurs.

While blooms in the North Sea can give increased concentrations of DMS and related compounds, blooms are thought to have less overall importance in the generation of DMS than do the general background levels in great areas of the oceans. Concentrations measured in coastal marine waters have a range of 1-10 nM DMS, 1-10 nM dissolved DMSP and 0-200 nM particulate DMSP.

A number of different microorganisms have been isolated, using DMS and DMSP as substrates, from the water column of oceans. They thus have high a affinity for DMS but may be very slow-growing, which may have relevance to the nmolar concentrations found in the oceans. Marine bacteria are calculated to have a turnover of one per day. Therefore, to achieve this mean turnover rate on such low substrate concentrations at least some of them have to be growing mixotrophically or using many different substrates simultaneously.

The marine environment has many organisms able to use acrylate and sarcosine, indicating the breakdown of DMS to acrylate to be important as a nutrient in the marine environment. Acrylate is difficult to measure, and not very much is detected in the marine environment because it is used very rapidly, although the biochemistry of the process is not known. DMSP breakdown by DMSP lyase to DMS and acrylate accounts for less than 50% of lost DMSP whereas DMSP conversion to methanethiol and 3-mercaptopropionate via two sequential demethylations, and the intermediate 3-methiolpropionate, is thought to account for more than 50% of DMSP loss. Therefore DMS production may previously have been overestimated because of earlier undervaluation of this demethylation pathway.

Methanethiol is very reactive and binds with surfaces and particles. It also occurs in fuels and can be a source of sulfur in coals. Studies of a freshwater system showed DMS was produced by methylation of methanethiol. Methylation of methanethiol also takes place in the marine environment but it is not known how important this is relative to the conversion of DMSP to DMS. Marine studies with DMSP demonstrated production of DMS and methanethiol but that the DMS persisted whereas methanethiol was lost or used.

Most work on methylated sulfur compounds to date seems to be in marine and freshwater environments such as sediments, but DMS has a number of known roles in a variety of environments. As with many other sulfur compounds, DMS may be an attractant and acts as a repellent of herbivores when produced by marine plants. It can also influence denitrification. DMS also arises from salt marshes and sphagnum bogs. In sphagnum bogs DMSO and S-methylmethionine gave rise to DMS in slurries, but DMSP did not. Syringic acid stimulated DMS production as did trimethoxybenzoate and methanethiol. It has been shown that DMS consumption is linked to methanogenesis at pH values greater than pH 4.6 but not at values below pH 4.6. DMS can be a precursor for methane in some areas and methane can be a sink for DMS in some peat bogs. In some methane-producing sediments, the principal fate of DMS can indeed be conversion to methane by methylotrophic methanogens.

Methylamine fluxes are uncertain, as is the environmental significance of their flux to the atmosphere. Marine environments are known to release monomethylamine and dimethylamine to the atmosphere but there are difficulties of measurement since some adsorb to glass equipment (but not to plastic).

Carbon disulfide (CS_2) and carbonyl sulfide (COS)

Information on biological COS production, breakdown and turnover is relatively sparse despite its importance as the most abundant sulfur gas in the atmosphere. Sulfate reducers have been reported to make a range of organosulfur compounds including COS. Dimethylsulfoniopropionate (DMSP) has been reported to produce COS when exposed to light. The oceans contain CS_2 and COS and have been reported as both sources and sinks of COS depending on variables such as pH and temperature. At Lake Giso in Spain there are increased emissions of Volatile Organic Sulfur Compounds during holomyxis. The stratified lake produces CS_2 below the epilimnion which decreases in concentration as it rises through the lake body. This is thought to be primarily a biochemical decrease rather than a chemical one. Salterns release COS, CS_2 and methanethiol, but COS dominates as the salt concentration increases. CS_2 degradation has been shown in some soil and swamp samples.

Areas of interest for study in relation to fluxes

To date, work on the biogeochemistry of organosulfur gases has been led by pure culture studies in which organisms have been sought and isolated the basis of biogeochemical observations. Much of the past and current research has been carried out using concentrations of these compounds that are very high relative to those encountered in nature, but the contribution to flux of both low and high concentrations is recognized as important. While processes are studied using environmental samples rather than pure cultures, cross-fertilization between process experiments and pure culture experiments needs to be enhanced to optimize the information output. The following topics for development of present studies in the future can be identified:

1. Ecophysiology of microorganisms in areas of low substrate concentration in both marine and fresh water environments

Metabolic potentials and different substrates and survival mechanisms are key parameters. There is a need to investigate affinities of pure cultures for target substrate compounds and to compare with organisms at natural substrate concentration levels. Competition for organic sulfur substrates and the role of affinity and competition in determining flux and degradation rates need further study. In addition to the free-living bacteria, the contribution of gut bacteria (of marine animals) may be of significance. Copepod guts are known to host methane generation and oxidation by bacteria.

2. The nature of the role of marine nitrifiers in oxidation of methane

It is known that nitrifying bacteria can convert DMS to DMSO and that some can obtain energy from formaldehyde oxidation. In the marine environment there is photochemical production of formaldehyde (10-100 nM/day), so formaldehyde has potential importance in maintaining nitrifier populations. Their contribution to methane oxidation is subject to considerable uncertainty.

3. More needs to be known of the biochemistry of the processes as occurring in natural habitats

While much is known about many of the biochemical pathways exhibited by well-studied organisms, there is a need to develop functional probes to determine the extent of activities and processes in natural environments. Experience has shown that antibody probes are unsatisfactory for work with environmental samples. Ultimately there is a need to make probes for these enzymes to see how important and widespread they are.

4. Metabolic diversity

In anaerobic sediments DMSP concentration can be much higher than in the water column and there may be different transport mechanisms used among these microbes. Organically-bound sulfur, rather than sulfate, is found in freshwater sediments. Sulfur gases such as DMS and thiols from freshwater sediments are not well studied.

5. Anaerobic processes

Another question of interest relating to anaerobic environments concerned the role of sulfur species in anaerobic methane oxidation proceeding via methanethiol, methyl-coenzyme M and dimethylthiolase. $^{14}CH_4$-labelling and autoradiography might be used to look at anaerobic methane oxidation but there are problems with incorporation because of the low free energy available thermodynamically. There may be a linkage of anaerobic methane oxidation to Fe or Mn species rather than sulfate in some areas where there are reactive iron species available, but addition of iron and/or manganese has not been tested to see if methane removal is stimulated.

6. Methodologies

Devising suitable procedures can be a problem in field work. One of the general problems identified in this type of research was that of safety and regulations concerning safe handling procedures, since many of the volatile sulfur cycle intermediates are toxic. Analysis and measurement is complex, as microelectrodes are unsuitable for use with many of the sulfur gases. DMSP lyase is a robust enzyme and could possibly be used as a sensor for measuring DMSP/DMS. It is very specific for DMSP.

7. Behaviour of organisms in natural environments and their response to environmental changes

Environmental changes such as in temperature or drying have an impact on the ability of these organisms and their metabolism. Circulation patterns in oceans can change with changes in climate which may have marked effects on bioprocesses.

8. Lacunae in knowledge

Most data on biological turnover of trace gases relates to the marine environment. There are tremendous gaps in knowledge of the terrestrial and freshwater systems. The contribution of extreme environments such as Antarctica (psychrophilic turnover) and the microlayer of oceans also requires clarification.

9. Microbial diversity and metabolic plasticity

A vast array of different bacteria and othewr microorganisms contribute to trace gas production and consumption, but the relationships of diversity and plasticity to turnover of trace gases needs much more study. The best methods to use are, however, debatable: mRNA probes? Enzyme probes? 16S rRNA only tells us "who" is there and not what they are doing - if anything. The presence of an organism type cannot necessarily be deemed to be significant unless it can be shown to be actively involved, directly or indirectly, in the dynamics of the process of interest. The target needs to be highly conserved (like a *nif* gene equivalent) to be able to detect organisms of interest with, for example, DMSP lyase.

Perspectives on the Microbiology of Atmospheric Trace Gases

Donovan P. Kelly
Institute of Education
University of Warwick
Coventry CV4 7AL
England

Uncertainties of identity and contribution

The Chapters and Working Group Reports in this volume serve to demonstrate both (i) the enormous scale of the biogeochemical processes involving trace gases and (ii) the difficulties of working *in situ* with the ambient concentrations of compounds commonly encountered. The other problem always in the minds of those looking at the biological turnover of trace gases is whether the organisms isolated and studied by them are actually those contributing in main or in part to that turnover. Examples are (i) the competition between methanotrophs and ammonia-oxidizing nitrifiers for the oxidation of methane, (ii) competition for the oxidation of dimethyl sulfide by chemolithotrophic thiobacilli and facultatively heterotrophic hyphomicrobia, (iii) the balance between aerobic nitrifiers and denitrification for the destruction of nitrous oxide, and (iv) the possibility that the destruction of carbonyl sulfide (COS) in the marine environment, may be primarily catalyzed by the CO_2-concentrating systems of phytoplankton, effecting hydrolysis to H_2S, rather than by chemolithotrophic thiobacilli. A parallel uncertainty can be whether organisms isolated by virtue of their capacity to use a particular compound are indeed able to use that compound at the concentrations actually available in the habitat. For example, some bacteria able to oxidize methane, carbon monoxide or dimethyl sulfide in pure culture exhibit affinities for those compounds (as K_m or K_s values) that are so poor that their rates of consumption of those materials, given the low concentrations actually prevailing in the source environment, may contribute little to the turnover of the

NATO ASI Series, Vol. I 39
Microbiology of Atmospheric Trace Gases
Edited by J. Colin Murrell and Donovan P. Kelly

compounds in Nature. For example, the detection of nanomolar concentrations of some substrates in natural waters may be accompanied by isolation of bacteria with K_s values in the μmolar range, casting doubt on their role in turnover *in situ*. Clearly, *in situ* demonstration of microbiologically-catalysed turnover indicates the existence of active bacteria, possibly of an "unculturable" nature. Of course, in many cases local concentrations of "rare" compounds (e.g. DMS, CS_2, NO) may reach concentrations quite high enough for their metabolism to be effected by known organisms already obtained in pure culture.

The importance of the development and use of specific probes for detecting the presence of enzyme systems identified from pure culture biochemical studies cannot be overemphasized: probes for methane monooxygenase, MSA monooxygenase and enzymes specific for the metabolism of other important trace gases (such as NO, DMS, COS and CS_2) are among the most important indicators needed for use in environments where these compounds undergo biological reaction, in order to determine whether the organisms involved are at least similar to those from which the probes have been derived.

The case of methanesulfonic acid (MSA)

The biodegradation of MSA is discussed in detail by Murrell et al. (this volume). This compound is possibly unique in being an example of a natural product with only one known biogenic source for input into the biosphere: the atmospheric oxidation of dimethyl sulfide. DMS is produced at an annual rate of 40-70 Tg, mainly from phytoplankton, with a mean of around 26 Tg DMS-sulfur per annum. Depending on the latitude, DMS is oxidized to various ratios of MSA and SO_2, but overall 25-70% of it is oxidized to MSA (Mihalopoulos et al. 1992; Kelly et al. 1993, 1994; Thompson et al. 1995). It is believed that MSA undergoes no significant further chemical reaction in the atmosphere so it must (under steady state conditions) be deposited at an essentially constant annual rate back on to the Earth in rain, snow and by dry deposition. This makes it possible to calculate the maximum concentrations of MSA that might deposit in rain on to soils (and of course into the oceans, resulting in immediate dilution of the input concentrations).

The generation of MSA in the atmosphere can be estimated from the data above to be in the range 19.5-54.6 Tg MSA per annum (Murrell et al. this volume; Kelly and Murrell 1996). The amount of water precipitating on the Earth annually in the hydrological cycle is put at 4×10^{17} kg. Making the crude (and inaccurate) assumption that the MSA is evenly distributed in all the rain and snow falling on the Earth, a maximum mean concentration can be calculated as:

$$54.6 \text{ Tg MSA} = 5.7 \times 10^{11} \text{ moles MSA}$$
$$4 \times 10^{17} \text{ kg } H_2O = 4 \times 10^{17} \text{ liters}$$
$$\text{Concentration of MSA} = [5.7/4.0] \times 10^{-6}\text{M} = 1.4\ \mu\text{M}$$

This value is obviously a considerable approximation, but given latitudinal variation in extent of MSA formation and consequently some local variation in the concentration of MSA, a natural global range on the scale 0.3-4 μM would not seem unrealistic. Given that the K_m values reported for bacteria able to grow on MSA in pure culture are around 4 μM, it seems likely that they might well be exposed to natural inputs of this substrate at useful concentrations in the natural environment.

Another factor suggesting that the MSA-degrading methylotroph strains so far isolated are indeed important in the natural environment is their possession of an oxygenase with a marked specificity for MSA (MSAMO). The purified MSAMO enzyme from methylotroph Strain M2 will only catalyze the consumption of NADH with three other substrates: ethane- and propane-sulfonic acids and monomethyl sulfate. In that these compounds are probably even less abundant in the environment than MSA, it suggests that the *in vivo* function of MSAMO must be for MSA oxygenation and not for an alternative metabolic function.

The problem of anaerobic sinks

Anaerobic environments as sources of trace gases are well documented, and in many cases the gas-producing bacteria they contain have been studied in great functional and taxonomic detail. Diverse anaerobic sources are known for CH_4, N_2O, DMS, H_2S, COS and CS_2 and other sulfur gases.

For some of the trace gases, no anaerobic sinks have been demonstrated, and for others there is doubt about the nature of the anaerobic destruction processes. For DMS, a considerable literature has developed on its anaerobic transformations, including conversion to methane by methylotrophic methanogens. In the case of some other gases, no organisms have been unequivocally demonstrated as capable of using them for anaerobic growth, and there is little or no functional evidence in the form of established enzyme systems or metabolic mechanisms. The enigma of methane is the best known, but anaerobic methane oxidation is now accepted as occurring in some situations, possibly catalysed by consortia involving sulfate-reducing bacteria. No anaerobic sinks for MSA or CS_2 are yet known, involving bacteria able to use these compounds as anaerobic energy and carbon substrates.

In the cases of each of CH_4, MSA and CS_2, the initial aerobic metabolic attack on them involves an oxygenase, with addition of molecular oxygen to the carbon of the molecules. Anaerobic metabolism must therefore be initiated by an alternative to direct oxygenation. For MSA and CS_2, hydrolysis can be postulated:

$$CH_3SO_3H + H_2O = CH_3OH + H_2SO_3 \quad [1]$$

$$CH_3SO_3H + H_2O = CH_4 + H_2SO_4 \quad [2]$$

$$CS_2 + H_2O = COS + H_2S \quad [3]$$

$$COS + H_2O = CO_2 + H_2S \quad [4]$$

Reaction [1] could support anaerobic growth on the methanol product, but reaction [2] would be a "dead-end" sink, producing the most reduced form of carbon and the most oxidized form of sulfur. Reactions [3] and [4] produce H_2S that could support growth of denitrifying sulfur-oxidizing chemolithotrophs. The free energy change (as ΔH or ΔF) of reaction [1] has not been calculated, as enthalpy values for MSA seem not to have been determined or calculated. For reactions [3] and [4], the free energies (ΔF^o) of the reactions can be calculated from the free energies of formation of the reactants (Latimer 1952). These are: [3] +91.4 kJ and [4] -29.6 kJ (Kelly, newly calculated). Reaction [3] is thus somewhat unfavourable energetically. It, and reaction [4], do seem to be catalysed by anaerobic suspensions of *Thiobacillus thioparus* (Smith and Kelly 1988). The simultaneous occurrence of reactions [3] and [4] may well favour the "pulling" of reaction [3] by the consumption of its product (COS) by reaction [4].

It is possible that MSA could act as a substrate anaerobically for either or both of methanogens and sulfate-reducing bacteria, being used as a respiratory hydrogen acceptor:

$$CH_3SO_3H + 4\ H_2 = CH_4 + H_2S + 3\ H_2O \qquad [5]$$

For CH_4, hydrolysis is a less obvious solution, and would presumably need coupling to a hydrogen acceptor. A hypothetical possibility (requiring thermodynamic as well as microbiological assessment) could be:

$$CH_4 + H_2O = CH_3OH + 2[H] \qquad [6]$$

$$2[H] + N_2O = H_2O + N_2 \qquad [7]$$

$$\text{Overall: } CH_4 + N_2O\ [+H_2O] = CH_3OH + N_2\ [+ H_2O] \qquad [8]$$

The ΔF^o for the overall reaction [8] is -20.1 kJ (Kelly, newly calculated), indicating its thermodynamic feasibility.

Climatic impacts of trace gases produced and/or consumed by microorganisms

Methane is now well known as a trace gas of great radiative importance, with few doubts about its effects, sources and sinks. N_2O is becoming better understood, and is of undoubted radiative importance in climate homeostasis and change. The central importance of DMS in the biogeochemical cycling of sulfur has been established beyond all doubt over the past twenty years, and is currently the subject of vast amounts of work.

The situation is less certain for other sulfur trace gases, and there is urgent need for continued evaluation of (i) the DMS/MSA connection, (ii) the microbiology and environmental effects of MSA, (iii) the microbiology and climatic effects of COS and CS_2, and of the volatile thiols.

The environmental importance of DMS and MSA extend beyond their role in sulfur cycling, and include the functioning of MSA molecules as cloud condensation nuclei, potentially affecting albedo, atmospheric temperature, and precipitation. Moreover, "acid rain" due to the atmospheric oxidation of DMS to MSA and sulfate was suggested as being contributory to environmental degradation of European forests caused by acid rain arising from industrial SO_2 (see Kelly and Smith 1990). In that example, DMS concentrations were postulated to be regionally very high as a consequence of phytoplankton blooms in the North Sea. Most interestingly a recent report has linked the possibility of Late Cretaceous acid rain with the extinction of many plants and animals in that period (Robinson 1995). This hypothesis takes account of the fact that during the Late Cretaceous period (duration 30.7 Ma, ending about 70 Ma before the present) the chalk deposits still extant today were deposited by coccolithophores in the Cretaceous chalk seas. It is thus likely that the coccolithophores produced abundant amounts of DMS, which was oxidized to MSA in the atmosphere. This could have produced regionally high MSA deposition, with total global deposition possibly greatly exceeding that in modern times. Climatic change resulting in stimulation of phytoplankton blooms (particularly of prymnesiophytes and other DMS-producers) could produce highly undesirable effects in the future, were it to occur.

If we assume a mean deposition of 50 Tg MSA/year over the past 100,000,000 years of geological time, the amount of MSA deposited into the biosphere must have been of the order of 5 x 10^{15} tonnes in that period. Presumably this, together with MSA undoubtedly deposited in earlier aeons, could well have been sufficient to have had significant long-term climatic and environmental effects. This period of time would also have been ample to have allowed microbial evolution of the use of this compound in diverse ways, perhaps most of which remain to be discovered.

References

Kelly DP, Murrell JC (1996) Metabolism of methanesulfonic acid. In Lidstrom ME, Tabita FR (eds) Microbial growth on C_1 compounds. Kluwer Netherlands (in press)
Kelly DP, Smith NA (1990) Organic sulfur compounds in the environment: biogeochemistry, microbiology, and ecological aspects. Adv Microb Ecol 11:345-385
Kelly DP, Malin G, Wood AP (1993) Microbial transformations and biogeochemical cycling of one-carbon substrates containing sulphur, nitrogen or halogens. In Murrell JC, Kelly DP (eds) Microbial growth on C_1 compounds, pp 47-63. Intercept Andover
Kelly DP, Baker SC, Trickett J, Davey M, Murrell JC (1994) Methanesulphonate utilization by a novel methylotrophic bacterium involves an unusual oxygenase. Microbiology (UK) 140:1419-1426
Latimer WM (1952) The oxidation states of the elements and their potentials in aqueous solutions. Prentice-Hall Englewood Cliffs NJ
Mihalopoulos N, Nguyen BC, Boissard C, Canpin JM, Putaud JP, Belviso S (1992) Field study of dimethylsulfide oxidation in the boundary layer: variations of dimethylsulfoxide, methanesulfonic acid, sulfur dioxide, non-seasalt sulfate and Aitken nuclei at a coastal site. J Atmos Chem 14:459-477
Thompson AS, Owens NJP, Murrell JC (1995) Isolation and characterization of methanesulfonic acid-degrading bacteria from the marine environment. Appl Environ Microbiol 61:2388-2393
Robinson ND (1995) Biogenic acid rain during the Late Cretaceous as a possible cause of extinctions. J Geol Soc 152:4-6
Smith NS, Kelly DP (1988) Oxidation of carbon disulphide as the sole source of energy for the autotrophic growth of *Thiobacillus thioparus* strain TK-m. J Gen Microbiol 134:3041-3048

Index

Acetogenic bacteria 91
Acetol 107
Acetone 106
Acetylene 153, 155, 176, 190, 258
Acid rain 294
Acrylic acid 208, 283
Adenosine phosphosulfate 228
S-Adenosylmethionine 229
Aerosols
- sulfate 205

Agrobacterium 251
Alcaligenes faecalis 176, 180
Alkylsulfonates 245
Ammonia monooxygenase 123, 128, 138, 141,
- genes 143-146, 176

Ammonia oxidising bacteria 25, 74
Anal sac secretions 103
Anaerobic sinks 291-293
Antibody techniques 138
Anthropogenic methane 33
Arrhenius function 42, 60
Atmospheric mixing ratios 87
ATP-sulfurylase 228
Azide 248

Bacillus thermoglucosidasius 104
Bering Sea 65
Biocides 94
Biofilms 156, 162
Biomass burning 34, 51
Biosensor
- methane 160-161

Bromoethanesulfonic acid 54
Bromoform 85
Bunsen solubility coefficient 74
Buried bag technique 72
Butanethiol 233

Carbon dioxide 3, 265
Carbon disulfide 227, 236, 259, 284, 292-293
Carbon monoxide 8, 73, 248, 258
Carbonyl sulfide 8, 227, 234, 284, 289, 292-293
- free energy change 292-293

Carnitine 209
Catabolite repression 104
Cathode
- silver 154
- gold 154

Chlorofluorocarbons 85,87-89, 259
Chloroform 214
Climate change 2, 13, 205, 294
Cloud condensation nuclei 243, 262
Clover leaf 156-157
Coal mining 51
Coastal marine sediment 64
Cometabolism 114
Copiotrophs 20
Copper 129-131, 137, 171, 182
Cyanide 248
Cyclopropanol 246

Denitrification 168, 257
Diauxie 116-117
Dibromomethane 85
Dicyandiamine 190
Diffusion coefficient 42
- reaction model 157

Dimethyl arsine 12
Dimethyldisulfide 227
Dimethylsulfide 8, 227, 259, 268, 289, 290
- cretaceous coccolithophores 294

Dimethylsulfoniopropionate 11, 207-211, 229, 243
- lyase 234

Dimethylsulfoxide 214, 233
Dimethylthiolase 286
DNA hybridisation 122-123
DNA polymerase 124

Electron-capture gas chromatography 88
Electrophoresis 274
Electron transport phosphorylation 170, 179
Ellman's reagent 248
Emiliana huxleyi 208, 243
Escherichia coli 169
Ester hydrolysis 111
Ethanethiol 215, 227
Ethanol 111, 114
Ethyl acetate 103, 109-114
Ethyl formate 107
Ethylmethyl sulfide 216

Facultative methylotroph 38
Farmland 70
Fatty acid analysis 138
Ferredoxin 249
Ferric oxyhyrdoxides 231
Fertilizer
effects 45
Flow cytometry 274
Fluorescein 124
Fluorescence microscopy 147, 251
Fluorescence antibodies 121-122
Fluoroform 96
Formaldehyde
assimilation 135, 245
Freshwater
sediments 58
Fusarium oxysporum 171

Gaussian distribution 43
Gene-probes 127, 131, 140, 290
Global methane budget 27, 51
Glycine betaine 209, 235
Grassland 43
Greenhouse effect 2, 7, 261

Harvard forest 72
Heme 182
Hole-in-the pipe-model 187-191
Homocysteine 234
Human breath 103
Hybridization 139
Hydrochlorofluorocarbons 85, 89-90
Hydrofluorocarbons 85, 89-90
Hydrogen sulfide 161, 292, 293
Hydroxylamine 169, 174
Hydroxyl radicals 86, 95
Hydroxypyruvate reductase 245
Hyphomicrobia 233-235

Ice cores 4
Isopropyl alcohol 107

Kattegat 56

Lake Constance 38, 59
Lake Washington 59
Landfill 33, 267
cover soil 37, 38, 45
Land-use 69, 81
Laser scanning confocal
microscopy 274

Liebig
Law of the limiting factor 40
Limfjorden 56

3-Mercaptopropionate 234
Methane
bubble 57
budget 64
hydrate 51
seeps 52
Methane monooxygenase 20, 22, 27, 38, 90, 123, 128, 136-152
genes 137-138, 141-145, 280
Methane oxidation
agricultural soil 69-84
ammonia inhibition 45-46
anaerobic 53-58, 66, 292, 293
field measurements 72, 75, 77
kinetics 39, 42, 129-132
peat 40, 140, 278
potential 22, 40
soil 71
temperature effects 61
temperate forest 69-84
Methane phosphonate 12
Methanesulfonic acid 6, 243-252, 290-294
monooxygenase 246-252
Methanethiol 211, 232, 283
Methanogenesis 13, 18, 42, 162, 218-219
Methanol dehydrogenase 139, 245
Methanotrophs
cysts 22, 26
high affinity 37
marine 21
molecular ecology 135-151
spores 22, 26
starvation 23-25
3-Methiolpropionate 210
Methionine 215, 234
Methyl acetate 109-114
Methylammonium 105, 114, 244, 284
Methyl bromide 91, 93-94, 233
Methyl chloride 8, 91
Methyl-coenzyme M 286
Methyl fluoride 89
Methyl formate 103, 107

Methyl halides 91-94
Methyl iodide 91, 233
S-Methylmethionine 214
Methylobacter albus 23, 135, 141
Methylobacterium 251
Methylococcus capsulatus 89, 128, 135, 137
Methylocystis 135, 141
Methylomonas 135, 143
Methylosinus trichosporium 22, 89, 129, 135, 137, 141
Methylotroph
    strain M 244-252
Microprofile
    methane 165
Microsensor
    construction 155
    oxygen 164
Mixotrophy 38, 116-117
Molecular ecology 138-147, 251, 272, 273
Molybdate 54
Monofluoroacetate 96
Montreal Protocol 86, 95
Mutagenesis 249

Naphthalene
    oxidation 22
Natural gas
    seeps 35-36
Nitrapyrin 176, 190
Nitrate reductase 170
Nitric oxide reductase 170
Nitrification 72, 81, 176
Nitrifiers 46, 93
Nitrite reductase 170, 171, 182
Nitrobacter 180
Nitrobacteria 175-177
Nitrogen
    fertilization 69, 70, 80
    mineralization rates 76
Nitrosobacteria 172-174
Nitrosococcus oceanus 143
Nitrosolobus multiformis 144
Nitrosomonas europaea 141, 169
Nitrosomonas eutropha 177
Nitrosospira 144
Nitrous oxide 293
    reductase 170
Non-sea salt sulfate 227
N-serve 73

Oligonucleotide probes 139, 250, 279
Oligotrophic
    bacteria 19, 28
Oxygen
    transport 62
Ozone 86

Paracoccus denitrificans 171
Paracoccus versutus 104, 106
Peroxyacetyl nitrate 178
Phaeocystis 208, 243
Pheromones 103
Phosphoadenosine-
    phosphosulfate 229
Phospholipid
    analysis 138
Photochemical destruction 9
Photosynthesis 13
Phylogenetic probes 123, 124
Phytogel 245
Plant roots 43-44
    gas transport 44
Plasmids
    broad-host-range 250
Polymerase chain reaction (PCR) 124, 140, 273
Polysulfide 231
Propellants 87
Propylene oxide 136
Proteobacteria 127, 139
Pseudomonas aeruginosa 170
Pseudomonas fluorescens 180
Pyrite
    oxidation 231
Pyrroloquinoline quinone 246

Radiative
    forcing 7
    heating effect 4
Redox potentials 177
Reductive dechlorination 88
Refrigerants 87
Restriction fragment length polymorphism 139
Rhizome 26
Rhizosphere 43
Rhodamine 124
Rhodanese 230
Ribulose bisphosphate
    carboxylase 136, 237, 245
Ribulose monophosphate
    pathway 135
Rice paddies 17, 20, 33, 87, 266, 268
Ripening fruits 103
Rothamsted 45, 146

16S ribosomal RNA (rRNA) 123, 125, 136, 141-151, 251, 274, 287

Sea-air exchange 207
Sedimentary carbonates 4
Sedimentation rate 57
Sediments
 marine 55,56
 organic-rich 58, 66
Serine
 pathway 135, 245, 246
Sewage 163
Single-cell protein 136
Spartina 243, 282
Sphagnum 205, 206, 213
Starvation response 21, 24, 25
Static chambers 72
Strawberry fields 91, 94
Sulfate-methane
 reduction zone 53
 transition zone 54, 55
Sulfate-reducing bacteria 34, 92
Sulfate-reduction
 rates 36
Sulfide oxidase 230
Sulfite oxidase 230
Sulfite reductase 229
Sulfur cycle 3
Sulfur oxidase 230
Symbiosis 19
Syringic acid 215

Termites 34, 87, 267
Tetrafluoroethane 85
Thiobacillus thioparus 234-236
Thiobacillus versutus 104
Thiocyanate 236
 hydrolase 234
Thiophenes 232
Thiosphaera pantotropha 171
Thiothrix ramosa 236
Transmethylation 11
Trichloroethylene 38, 136
Trifluoroacetate 95-97
Tropopause 86
Troposphere 86
Tundra 43

Vertical transport
 gases 43
Water table 40, 268
Wetlands 17, 265
 northern 18
 tropical 18

Participants at Workshop

Stef van Bergeijk
Laboratory for Microbiology
University of Amsterdam
Nieuwe Achtergracht 127
THE NETHERLANDS

Levente Bodrossy
Institute of Biophysics
Biological Research Centre
Hungarian Academy of Sciences
PO Box 521, H-6701 Szeged
HUNGARY

Oswald van Cleemput
Faculty Agricultural and Applied Biological Sciences
University of Ghent
Coupure 653, B-9000 Ghent
BELGIUM

Ralf Conrad
Max-Planck Institute for Terrestrial Microbiology
Karl-von-Frisch Str.
D-3350 Marburg
GERMANY

Andria M Costello
Environmental Engineering Science
W M Keck Laboratories 138-78
California Institute of Technology
Pasadena, CA 91125
USA

Lars Damgaard
Institute of Biological Sciences
University of Aarhus
B. 540, Ny Munkegade, DK-8000 Aarhus C
DENMARK

Rosa de La Torre
Division of Space Research
I.N.T.A.
28850-Torrejon de Ardoz
SPAIN

Peter Frenzel
Max-Planck-Institute for Terrestrial Microbiology
Karl-von-Frisch-Str.
D-35043 Marburg
GERMANY

Theo A Hansen
Department of Microbiology
University of Groningen
Kerklaan 30
NL-9751, NN Haren
THE NETHERLANDS

Kaj Henriksen
Environmental Engineering Laboratory
Aalborg University
Sohngaardsholmsvej 57
DK-9000 Aalborg
DENMARK

Andrew Holmes
Department of Biological Sciences
University of Warwick
Coventry CV4 7AL, UK

Niels Iversen
Environmental Engineering Laboratory
Aalborg University
Sohngaardsholmsvej 57
DK 9000, Aalborg
DENMARK

Sigmund Jensen
Agricultural University of Norway
Department of Biotechnical Sciences
PO Box 5040, N-1432 As
NORWAY

Sarah Jordan
Division of Life Sciences
Kings College London
Campden Hill
London W8 7AH, UK

Samantha Joye
Center for Environmental Studies
San Fransisco State University
3150 Paradise Drive
PO Box 855
Tiburon
CA 94920
USA

Hariklia Kargiolaki
Chania GR-73131
GREECE

Don P Kelly
Institute of Education
University of Warwick
Coventry
CV4 7AL, UK

Elizabeth Kenna
Department of Biological Sciences
University of Warwick
Coventry
CV4 7AL, UK

Ron P Kiene
Department of Marine Sciences
University of South Alabama
LSCB 25
Mobile, AL 36688
USA

Gary King
Darling Marine Center
University of Maine
Walpole
Maine 04573
USA

Asa Kasimir Klemedtsson
Swedish Environmental Research Institute
PO Box 47086
S-402 58 Göteborg
SWEDEN

Charlotte Kruse
Environmental Engineering Laboratory
Aalborg University
Sohngaardsholmsvej 57
DK-9000 Aalborg
DENMARK

Nilgun Kubilay
Middle East Technical University (METU)
Institute of Marine Sciences (IMS)
PO Box 28 Erdemli, 33731
TURKEY

Mary E Lidstrom
Environmental Engineering Science
W M Keck Laboratories 138-78
California Institute of Technology
Pasadena
CA 91125
USA

Werner Liesack
Max-Planck-Institute for Terrestrial Microbiology
Karl-von-Frisch-Str.
D-35043 Marburg
GERMANY

David Lloyd
Microbiology Group (PABIO)
University of Wales College of Cardiff
PO Box 915, Cardiff CF2 1TL
UK

Ian McDonald
Department of Biological Sciences
University of Warwick
Coventry CV4 7AL, UK

Gill Malin
School of Environmental Sciences
University of East Anglia
Norwich NR4 7TJ, UK

Pertti J Martikainen
Department of Environmental Microbiology
National Public Health Institute
P.O.B. 95
FIN-70701 Kuopio
FINLAND

J Patrick Megonigal
Department of Botany
Box 90339
Duke University
Durham, NC
USA

J Colin Murrell
Department of Biological Sciences
University of Warwick
Coventry
CV4 7AL, UK

David B Nedwell
Department of Biology
University of Essex
Wivenhoe Park
Colchester
CO4 3SQ, UK

Allan K Nielsen
Department of Molecular Biology
Odense University
DK-5230, Odense M
DENMARK

Ron S Oremland
US Geological Survey ms465
345 Middlefield Road
Menlo Park
California 94025
USA

Nikolai Panikov
Institute of Microbiology
Russian Academy of Sciences
Prospect 60-Let Octyabrya 7/2
Moscow, 117811
RUSSIA

Niels Peter Revsbech
Institute of Biological Sciences
Department of Microbial Ecology
University of Aarhus
Building 540, Ny Munkegade
DK-8000, Aarhus C
DENMARK

Peter Roslev
Environmental Engineering Laboratory
Sohngaardsholmsvej 57
DK-9000 Aalborg
DENMARK

J C M Scholten
Agricultural University
Department of Microbiology
H. van Suchtelenweg 4
6703 CT Wageningen
THE NETHERLANDS

Reinoud Segers
Department of Theoretical Production Ecology
Wageningen Agricultural University
P.O. Box 430, 6700 AK, Wageningen
THE NETHERLANDS

Rafel Simo
Department of Environmental Chemistry, CID-CSIC
Jordigirona 18-26
0834-Barcelona, Catalonia
SPAIN

Martin Slayne
Microbiology, ADAS Cambridge
Brooklands Avenue
Cambridge
CB2 2BL, UK

David R Smart
Centre de Recerce Ecologica i Applicacions Forestal (CREAF)
Universitat Autonoma de Barcelona
08193 Bellaterra, Barcelona
SPAIN

Jan Sorensen
Department of Ecology & Molecular Biology
Royal Veterinary and Agricultural University
Rolighedsvej 21, DK-1958 Frederiksberg C
Copenhagen
DENMARK

Paul A Steudler
Ecosystem Center
Marine Biological Laboratory
Woods Hole
MA 02543
USA

Bo H Svensson
Department of Water and Environmental Studies
University of Linkoping
S-583 81 Linkoping
SWEDEN

Pieter T Visscher
Department of Marine Sciences
University of Connecticut
1084 Shennescossett Road
Groton, CT 06340
USA

Andrea Watson
Department of Biology
University of Essex
Wivenhoe Park
Colchester, Essex
CM1 1QE, UK

Toby Willison
Soil Science Department, IACR Rothamstead
Harpenden, Hertfordshire
AL5 2JQ, UK

Ann Wood
Division of Life Sciences
King's College London
Campden Hill Road
London W8 7AH, UK

The ASI Series Books Published as a Result of Activities of the Special Programme on Global Environmental Change

This book contains the proceedings of a NATO Advanced Research Workshop held within the activities of the NATO Special Programme on Global Environmental Change, which started in 1991 under the auspices of the NATO Science Committee.

The volumes published as a result of the activities of the Special Programme are:

Vol. 1: **Global Environmental Change.**
Edited by R. W. Corell and P. A. Anderson. 1991.
Vol. 2: **The Last Deglaciation: Absolute and Radiocarbon Chronologies.**
Edited by E. Bard and W. S. Broecker. 1992.
Vol. 3: **Start of a Glacial.** Edited by G. J. Kukla and E. Went. 1992.
Vol. 4: **Interactions of C, N, P and S Biogeochemical Cycles and Global Change.**
Edited by R. Wollast, F. T. Mackenzie and L. Chou. 1993.
Vol. 5: **Energy and Water Cycles in the Climate System.**
Edited by E. Raschke and D. Jacob. 1993.
Vol. 6: **Prediction of Interannual Climate Variations.**
Edited by J. Shukla. 1993.
Vol. 7: **The Tropospheric Chemistry of Ozone in the Polar Regions.**
Edited by H. Niki and K. H. Becker. 1993.
Vol. 8: **The Role of the Stratosphere in Global Change.**
Edited by M.-L. Chanin. 1993.
Vol. 9: **High Spectral Resolution Infrared Remote Sensing for Earth's Weather and Climate Studies.**
Edited by A. Chedin, M.T. Chahine and N.A. Scott. 1993.
Vol. 10: **Towards a Model of Ocean Biogeochemical Processes.**
Edited by G. T. Evans and M.J. R. Fasham. 1993.
Vol. 11: **Modelling Oceanic Climate Interactions.**
Edited by J. Willebrand and D.L.T. Anderson. 1993.
Vol. 12: **Ice in the Climate System.** Edited by W. Richard Peltier. 1993.
Vol. 13: **Atmospheric Methane: Sources, Sinks, and Role in Global Change.**
Edited by M. A. K. Khalil. 1993.
Vol. 14: **The Role of Regional Organizations in the Context of Climate Change.**
Edited by M. H. Glantz. 1993.
Vol. 15: **The Global Carbon Cycle.**
Edited by M. Heimann. 1993.
Vol. 16: **Interacting Stresses on Plants in a Changing Climate.**
Edited by M. B. Jackson and C. R. Black. 1993.
Vol. 17: **Carbon Cycling in the Glacial Ocean: Constraints on the Ocean's Role in Global Change.**
Edited by R. Zahn, T. F. Pedersen, M. A. Kaminski and L. Labeyrie. 1994.
Vol. 18: **Stratospheric Ozone Depletion/UV-B Radiation in the Biosphere.**
Edited by R. H. Biggs and M. E. B. Joyner. 1994.
Vol. 19: **Data Assimilation: Tools for Modelling the Ocean in a Global Change Perspective.**
Edited by P. O. Brasseur and J. Nihoul. 1994.

Vol. 20: **Biodiversity, Temperate Ecosystems, and Global Change.**
Edited by T. J. B. Boyle and C. E. B. Boyle. 1994.

Vol. 21: **Low-Temperature Chemistry of the Atmosphere.**
Edited by G. K. Moortgat, A. J. Barnes, G. Le Bras and J. R. Sodeau. 1994.

Vol. 22: **Long-Term Climatic Variations – Data and Modelling.**
Edited by J.-C. Duplessy and M.-T. Spyridakis. 1994.

Vol. 23: **Soil Responses to Climate Change.**
Edited by M. D. A. Rounsevell and P. J. Loveland. 1994.

Vol. 24: **Remote Sensing and Global Climate Change.**
Edited by R. A. Vaughan and A. P. Cracknell. 1994.

Vol. 25: **The Solar Engine and Its Influence on Terrestrial Atmosphere and Climate.**
Edited by E. Nesme-Ribes. 1994.

Vol. 26: **Global Precipitations and Climate Change.**
Edited by M. Desbois and F. Désalmand. 1994.

Vol. 27: **Cenozoic Plants and Climates of the Arctic.**
Edited by M. C. Boulter and H. C. Fisher. 1994.

Vol. 28: **Evaluating and Monitoring the Health of Large-Scale Ecosystems.**
Edited by D. J. Rapport, C. L. Gaudet and P. Calow. 1995.

Vol. 29: **Global Environmental Change Science: Education and Training.**
Edited by D. J. Waddington. 1995.

Vol. 30: **Ice Core Studies of Global Biogeochemical Cycles.**
Edited by R. J. Delmas. 1995.

Vol. 31: **The Role of Water and the Hydrological Cycle in Global Change.**
Edited by H. R. Oliver and S. A. Oliver. 1995.

Vol. 32: **Atmospheric Ozone as a Climate Gas.**
Edited by W.-C. Wang and I. S. A. Isaksen. 1995.

Vol. 33: **Carbon Sequestration in the Biosphere.**
Edited by M. A. Beran. 1995.

Vol. 34: **Climate Sensitivity to Radiative Perturbations: Physical Mechanisms and Their Validation.**
Edited by H. Le Treut. 1996

Vol. 35: **Clouds, Chemistry and Climate.**
Edited by P. J. Crutzen and V. Ramanathan. 1996.

Vol. 36: **Diachronic Climatic Impacts on Water Resources.**
Edited by A. N. Angelakis and A. S. Issar. 1996.

Vol. 37: **Climate Change and World Food Security.**
Edited by T. E. Downing. 1996.

Vol. 38: **Evaluation of Soil Organic Matter Models.**
Edited by D. S. Powlson, P. Smith and J. U. Smith. 1996.

Vol. 39: **Microbiology of Atmospheric Trace Gases.**
Edited by J. C. Murrell and D. P. Kelly. 1996.

Vol. 40: **Forest Ecosystems, Forest Management and the Global Carbon Cycle.**
Edited by M. J. Apps and D. T. Price. 1996.

Vol. 41: **Climate Variations and Forcing Mechanisms of the Last 2000 Years.**
Edited by P. D. Jones, R. S. Bradley and J. Jouzel. 1996

Springer-Verlag and the Environment

We at Springer-Verlag firmly believe that an international science publisher has a special obligation to the environment, and our corporate policies consistently reflect this conviction.

We also expect our business partners – paper mills, printers, packaging manufacturers, etc. – to commit themselves to using environmentally friendly materials and production processes.

The paper in this book is made from low- or no-chlorine pulp and is acid free, in conformance with international standards for paper permanency.

MIX
Papier aus verantwortungsvollen Quellen
Paper from responsible sources
FSC® C105338

If you have any concerns about our products,
you can contact us on
ProductSafety@springernature.com

In case Publisher is established outside the EU,
the EU authorized representative is:
Springer Nature Customer Service Center GmbH
Europaplatz 3, 69115 Heidelberg, Germany

Printed by Libri Plureos GmbH
in Hamburg, Germany